ASAP

Environmental Science

By the Staff of The Princeton Review

princetonreview.com

Penguin
Random
House

The Princeton Review
110 East 42nd Street, 7th Floor
New York, NY 10017
Email: editorialsupport@review.com

Published in the United States by Penguin
Random House LLC, New York, and in Canada
by Random House of Canada, a division of
Penguin Random House Ltd., Toronto.

Terms of Service: The Princeton Review Online
Companion Tools ("Student Tools") for retail
books are available for only the two most recent
editions of that book. Student Tools may be
activated only once per eligible book purchased
for two consecutive 12-month periods, for a
total of 24 months of access. Activation of
Student Tools more than once per book is in
direct violation of these Terms of Service and
may result in discontinuation of access to
Student Tools Services.

ISBN: 978-0-525-56768-4
eBook ISBN: 978-0-525-56779-0
ISSN: 2576-5418

AP is a trademark registered and owned by the
College Board, which is not affiliated with, and
does not endorse, this product.

The Princeton Review is not affiliated with
Princeton University.

Editor: Colleen Day
Production Editors: Jim Melloan and Liz Dacey
Production Artist: Deborah A. Weber
Content Contributors: Jes Adams,
    Eliz Markowitz, and Katie Chamberlain

Printed in the United States of America.

10 9 8 7 6 5 4 3 2 1

Editorial

Rob Franek, Editor-in-Chief
Mary Beth Garrick, Executive Director of Production
Craig Patches, Production Design Manager
Selena Coppock, Managing Editor
Meave Shelton, Senior Editor
Colleen Day, Editor
Sarah Litt, Editor
Aaron Riccio, Editor
Orion McBean, Associate Editor

Penguin Random House Publishing Team

Tom Russell, VP, Publisher
Alison Stoltzfus, Publishing Director
Amanda Yee, Associate Managing Editor
Ellen Reed, Production Manager
Suzanne Lee, Designer

Acknowledgments

Jes Adams would like to thank Lauren Beck, Jennifer Vulanovic, Joanna Yang, and Neil Adams for all their help, feedback, and suggestions.

Eliz Markowitz would like to thank her friends, amazing father, and super awesome cat, Ezra, for their constant encouragement and support.

The Princeton Review would like to extend special thanks to Jes Adams, Eliz Markowitz, and Katie Chamberlain for lending their expertise and brilliant ideas to this project.

We are also indebted to our outstanding production artists, Debbie Weber and Craig Patches, for their stellar design, imagination, and endless patience and hard work. Thanks also to our production editors, Jim Melloan and Liz Dacey, for their time and attention to each page.

Contents

Get More (Free) Content

1 Go to **PrincetonReview.com/cracking**

2 Enter the following ISBN for your book: 9780525567684

3 Answer a few simple questions to set up an exclusive Princeton Review account. (If you already have one, you can just log in.)

4 Click the "Account Home" button, also found under "My Account" from the top toolbar. You're all set to access your bonus content!

Need to report a potential **content** issue?

Contact **EditorialSupport@review.com**.
Include:
- full title of the book
- ISBN
- page number

Need to report a **technical** issue?

Contact **TPRStudentTech@review.com**
and provide:
- your full name
- email address used to register the book
- full book title and ISBN
- computer OS (Mac/PC) and browser (Firefox, Safari, etc.)

Once you've registered, you can...

- Access a variety of printable resources, including bonus "could know" material and key information about AP Environmental Science labs

- Get our take on any recent or pending updates to the AP Environmental Science Exam

- Get valuable advice about the college application process, including tips for writing a great essay and where to apply for financial aid

- If you're still choosing between colleges, use our searchable rankings of *The Best 384 Colleges* to find out more information about your dream school

- Check to see if there have been any corrections or updates to this edition

Introduction

What Is This Book and When Should I Use It?

Welcome to *ASAP Environmental Science,* your quick-review study guide for the AP Exam written by the staff of The Princeton Review. This is a brand-new series custom built for crammers, visual learners, and any student doing high-level AP concept review. As you read through this book, you will notice that there aren't any practice tests, end-of-chapter drills, or multiple-choice questions. There's also very little test-taking strategy presented in here. Both of those things (practice and strategy) can be found in The Princeton Review's other top-notch AP series—*Cracking.* So if you need a deep dive into AP Environmental Science, check out *Cracking the AP Environmental Science Exam* at your local bookstore.

ASAP Environmental Science is our fast track to understanding the material—like a fantastic set of class notes. We present the most important information that you MUST know (or should know or could know—more on that later) in visually friendly formats such as charts, graphs, and maps, and we even threw a few fun facts in there to keep things interesting.

Use this book any time you want—it's never too late to do some studying (nor is it ever too early). It's small, so you can take it with you anywhere and crack it open while you're waiting for soccer practice to start, or for your friend to meet you for a study date, or waiting for the library to open.* *ASAP Environmental Science* is the perfect study guide for students who need high-level review in addition to their regular review and also for students who perhaps need to cram pre-exam. Whatever you need it for, you'll find no judgment here!

Because you camp out in front of the library like they are selling concert tickets in there, right? Only kidding.

Who Is This Book For?

This book is for YOU! No matter what kind of student you are, this book is the right one for you. How do you know what kind of student you are? Follow this handy chart to find out!

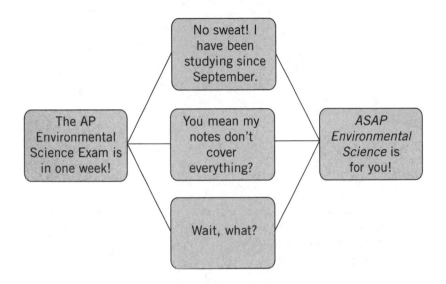

As you can see, this book is meant for every kind of student. Our quick lessons let you focus on the topics you must know, you should know, and you could know—that way, even if the test is tomorrow (!), you can get a little extra study time in, and only learn the material you need.

How Do I Use This Book?

This book is your study tool, so feel free to customize it in whatever way makes the most sense to you, given your available time to prepare. Here are some suggestions:

Target Practice

If you know what topics give you the most trouble, hone in on those chapters or sections.

ASK Away

Answer all of the ASK questions *first*. This will help you to identify any additional tough spots that may need special attention.

Three-Pass System

Start at the very beginning!*
Read the book several times from cover to cover, focusing selectively on the MUST content for your first pass, the SHOULD content for your second pass, and finally, the COULD content.

 It's a very good place to start.

ASAP Environmental Science

Why Are There Icons?

Your standard AP course is designed to be equivalent to a college-level class, and as such, the amount of material that's covered may seem overwhelming. It's certainly admirable to want to learn everything—these are, after all, fascinating subjects. But every student's course load, to say nothing of his or her life, is different, and there isn't always time to memorize every last fact.

To that end, *ASAP Environmental Science* doesn't just distill the key information into bite-sized chunks and memorable tables and figures. This book also breaks down the material into three major types of content:

 This symbol calls out a section that has MUST KNOW information. This is the core content that is either the most likely to appear in some format on the test or is foundational knowledge that's needed to make sense of other highly tested topics.

This symbol refers to SHOULD KNOW material. This is either content that has been tested in some form before (but not as frequently) or which will help you to deepen your understanding of the surrounding topics. If you're pressed for time, you might just want to skim it, and read only those sections that you feel particularly unfamiliar with.

This symbol indicates COULD KNOW material, but don't just write it off! This material is still within the AP's expansive curriculum, so if you're aiming for a perfect 5, you'll still want to know all of this. That said, this is the information that is least likely to be directly tested, so if the test is just around the corner, you should probably save this material for last.

As you work through the book, you'll also notice a few other types of icons.

 The Ask Yourself question is an opportunity to solidify your understanding of the material you've just read. It's also a great way to take these concepts outside of the book and make the sort of real-world connections that you'll need in order to answer the free-response questions on the AP Exam.

The Remember symbol indicates certain facts that you should keep in mind as you're going through the different sections.

 There's a reason why people say that "All work and no play" is a bad thing. These jokes help to shake your brain up a bit and keep it from just glazing over all of the content—they're a bit like mental speed bumps, there to keep you from going too fast for your own good.

There's a lot to think about in this book, and when you see this guy, know that the information that follows is always good to have on hand. You'll rock it in trivia, if no place else.

Where Can I Find Other Resources?

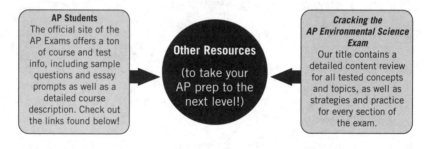

AP Students
The official site of the AP Exams offers a ton of course and test info, including sample questions and essay prompts as well as a detailed course description. Check out the links found below!

Other Resources
(to take your AP prep to the next level!)

Cracking the AP Environmental Science Exam
Our title contains a detailed content review for all tested concepts and topics, as well as strategies and practice for every section of the exam.

Useful Links

- AP Environmental Science Homepage: https://apstudent.collegeboard.org/apcourse/ap-environmental-science
- Your Student Tools: www.PrincetonReview.com/cracking
  See the "Get More (Free) Content" page for step-by-step instructions for registering your book and accessing more materials to boost your test prep.

CHAPTER 1

Earth Systems and Resources

AP Environmental Science requires an understanding of Earth's structure and the materials from which it is made. This chapter will review four of the physical spheres that make up our planet and regulate life on Earth: the lithosphere, pedosphere, atmosphere, and hydrosphere. Additionally, the biosphere (comprised of all living organisms) inhabits the planet and draws physical resources from the other four spheres. All five spheres interact to shape the variety of landforms, biomes, and phenomena that make environmental science so interesting.

Welcome to Planet Earth ⊚

Let's start by reviewing some earth science concepts.

Where Is Earth in the Solar System? ⊚

Earth is the third planet from the Sun in our solar system, which contains a total of eight currently known and recognized planets. From the Sun outward, these eight known planets are Mercury, Venus, Earth, Mars, Jupiter, Saturn, Uranus, and Neptune. Each planet has its own orbit in the shape of an ellipse (a "stretched" circle). It takes Earth about 365¼ days to complete its orbit around the Sun, a period of time we call a year.

 Did You Know?

Pluto was "demoted" from planetary status in 2006. However, astronomers theorize the existence of a "Planet Nine" yet to be discovered, based on gravitational phenomena in the outer reaches of the solar system.

Geologic Time Scale 💬

Earth is thought to be between 4.5 and 4.8 billion years old. The following **geologic time scale** will help you get a sense of the vast amount of time that has gone by since Earth was formed. You will not be responsible for memorizing all of the eons, eras, periods, and epochs for this exam, but you should be familiar with the major ones.

TIME UNITS OF THE GEOLOGIC TIME SCALE (Numbers are absolute dates in millions of years before the present.)				TIME RANGE OF SEVERAL GROUPS OF PLANTS AND ANIMALS
Eon	Era	Period	Epoch	

Eon	Era	Period		Epoch	(dates)
Phanerozoic Eon	Cenozoic Era	Quaternary		Holocene	
				Pleistocene	2
		Tertiary	Neogene	Pliocene	5
				Miocene	24
			Paleogene	Oligocene	37
				Eocene	58
				Paleocene	
					66
	Mesozoic Era	Cretaceous			144
		Jurassic			208
		Triassic			245
	Paleozoic Era	Permian			286
		Pennsylvanian			320
		Mississippian			360
		Devonian			408
		Silurian			438
		Ordovician			505
		Cambrian			570
Proterozoic Eon	Late				900
	Middle				1,600
	Early				2,500
Archeon Eon	Late				3,000
	Middle				3,400
	Early				3,800
Hadean	No record				

Precambrian comprises about 87% of the geologic time scale

Groups of plants and animals: Invertebrates, Fishes, Land plants, Amphibians, Reptiles, Mammals, Birds

Origin of Earth about 4.6 billion years ago

- We are currently in the Holocene epoch.
- Quaternary and tertiary are the two most recent geologic periods.
- Non-avian dinosaurs lived during the Mesozoic Era.

- Precambrian eons represent the vast majority of the geologic time scale.
- The next epoch will be called Anthropocene, to recognize human-kind's accelerating effects on the planet's physical resources, climate, and life forms. Many geologists propose that we have already entered this new, post-Holocene epoch.

Earth's Materials and Layers 💬

Planet Earth is made up of three concentric zones of rocks that are either solid or liquid (molten): the **core**, the **mantle**, and the **lithosphere**. Earth's layers can be defined by their chemical or physical properties.

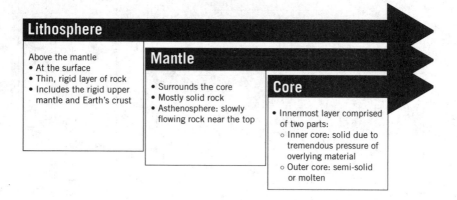

Lithosphere

Above the mantle
- At the surface
- Thin, rigid layer of rock
- Includes the rigid upper mantle and Earth's crust

Mantle

- Surrounds the core
- Mostly solid rock
- Asthenosphere: slowly flowing rock near the top

Core

- Innermost layer comprised of two parts:
  - Inner core: solid due to tremendous pressure of overlying material
  - Outer core: semi-solid or molten

The following diagram shows the chemical and physical properties of Earth's layers.

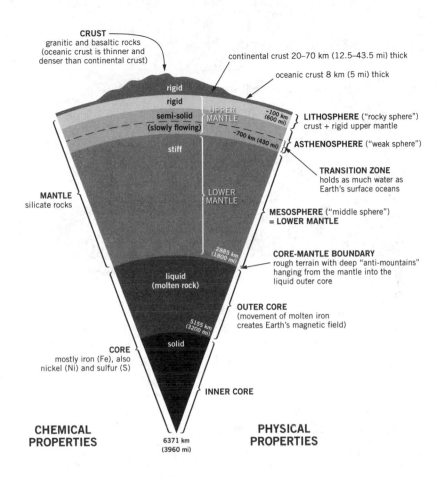

CRUST
granitic and basaltic rocks
(oceanic crust is thinner and
denser than continental crust)

continental crust 20–70 km (12.5–43.5 mi) thick

oceanic crust 8 km (5 mi) thick

rigid

rigid

UPPER MANTLE

semi-solid
(slowly flowing)

~100 km
(600 mi)

LITHOSPHERE ("rocky sphere")
crust + rigid upper mantle

~700 km (430 mi)

ASTHENOSPHERE ("weak sphere")

stiff

TRANSITION ZONE
holds as much water as
Earth's surface oceans

LOWER MANTLE

MANTLE
silicate rocks

MESOSPHERE ("middle sphere")
= **LOWER MANTLE**

2885 km
(1800 mi)

CORE-MANTLE BOUNDARY
rough terrain with deep "anti-mountains"
hanging from the mantle into the
liquid outer core

liquid
(molten rock)

OUTER CORE
(movement of molten iron
creates Earth's magnetic field)

5155 km
(3200 mi)

CORE
mostly iron (Fe), also
nickel (Ni) and sulfur (S)

solid

INNER CORE

**CHEMICAL
PROPERTIES**

6371 km
(3960 mi)

**PHYSICAL
PROPERTIES**

Note that the distances here indicate depth from the surface.

The Lithosphere ❗

The **lithosphere** is Earth's solid, rocky outer shell.

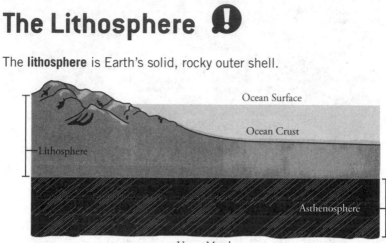

Ocean Surface

Ocean Crust

Lithosphere

Asthenosphere

Upper Mantle

diagram not to scale

Because the lithosphere floats atop the asthenosphere*, it can move around and break into large pieces. These pieces are called **tectonic plates.** Once separated, they move independently but also influence each other, and their movement may be linear or pivoting.

Earth's Tectonic Plates ❗

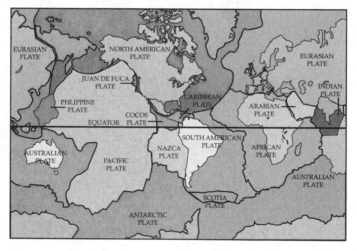

EURASIAN PLATE

NORTH AMERICAN PLATE

EURASIAN PLATE

JUAN DE FUCA PLATE

INDIAN PLATE

PHILIPPINE PLATE

CARIBBEAN PLATE

ARABIAN PLATE

COCOS PLATE

EQUATOR

SOUTH AMERICAN PLATE

AFRICAN PLATE

AUSTRALIAN PLATE

NAZCA PLATE

PACIFIC PLATE

AUSTRALIAN PLATE

SCOTIA PLATE

ANTARCTIC PLATE

😎 Think of this as like a cracker on top of a thick layer of hot pudding.

ASAP Environmental Science

Key Facts About Tectonic Plates 〰

- Some tectonic plates consist mostly of ocean floor.
  - Example: Nacza plate, a minor plate lying off the west coast of South America
- Other tectonic plates encompass both continental and oceanic crust.
  - Example: North American plate, where the United States is situated, which extends out to the mid-Atlantic ridge
- The only state in the U.S. without some portion on the North American plate is Hawaii, which lies completely on the Pacific plate.
- Pacific is the largest plate, occupying most of the Pacific Ocean basin along with southwestern California and Mexico's Baja Peninsula.

Plates Collide ❗

The edges of tectonic plates are called **plate boundaries,** and significant geological events occur where these boundaries abut each other. There are three types of plate boundary interactions.

1. Convergent Boundary	Two plates are pushed toward and into each other. One of the plates slides beneath the other, pushed deep into the mantle.
2. Divergent Boundary	Two plates move away from each other. This creates a gap between plates that may be filled with rising magma (molten rock). When this magma cools, it forms new crust.
3. Transform Fault Boundary	Two plates slide against each other in opposite directions (similar to rubbing your hands back and forth to warm them up).

What Happens When Plates Collide at Convergent Boundaries? !

The outcome of convergent boundaries depends on whether the collision happens between two oceanic boundaries, two continental boundaries, or at an oceanic-continental boundary.

Ocean-Ocean Convergence

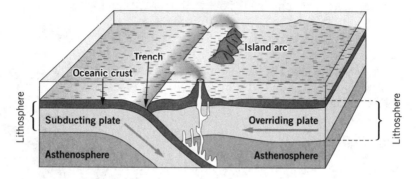

- **Subduction:** Older and denser plate sinks beneath the younger and lighter plate.
- A deep ocean trench forms at the point of subduction.
- An arc of volcanic islands may form on the side of the overriding plate.
- The subducting plate is melted into the mantle and recycled.

Continent-Continent Convergence

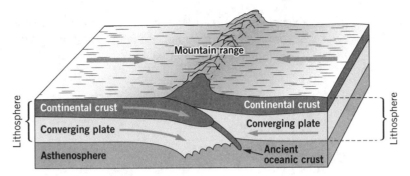

- No subduction because both plates are low density.
- This results in mountain building (orogeny).
- Earthquake activity is common.
- Volcanic activity is not common.

There is also **ocean-continent convergence,** not pictured here, which is similar to ocean-ocean convergence. In this scenario,

- the oceanic plate is subducted beneath the less dense continental plate, forming a deep ocean trench
- the subducting plate is recycled into the mantle, and volcanic mountains form on the continental side

---

 Did You Know?

- The Cascade Range is an example of ocean-continent convergence: the Pacific plate subducted beneath the North American Plate.
- The Himalayan Mountains were formed from the collision of the Indo-Australian plate with the Eurasian plate, a process that is still continuing. This is an example of continent-continent convergence.

Earthquakes ❗

Earthquakes are geological events resulting from vibrations deep in the earth that release energy. These vibrations are often due to sudden movements of Earth's tectonic plates sliding past one another at a transform boundary.

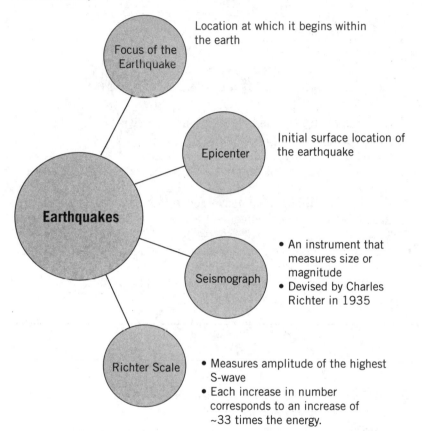

Focus of the Earthquake — Location at which it begins within the earth

Epicenter — Initial surface location of the earthquake

Earthquakes

Seismograph
- An instrument that measures size or magnitude
- Devised by Charles Richter in 1935

Richter Scale
- Measures amplitude of the highest S-wave
- Each increase in number corresponds to an increase of ~33 times the energy.

What did the ground say to the earthquake? You crack me up!

Volcanism ❗

Volcanoes are also geological events resulting from plate movement. Volcanoes are actually mountains formed by pressure from magma in Earth's interior, which rises through cracks or weaknesses in the crust. Pressure builds until the magma explodes through Earth's surface in a volcanic eruption.

Volcano Classification ❗

- **Active:** Erupted within the last 10,000 years
- **Dormant:** Has not erupted within the last 10,000 years but is expected to erupt again
- **Extinct:** Is not expected to erupt again

 Did You Know?

Sometimes a volcano believed to be extinct erupts. Two examples of this are the eruption of Mount Pinatubo in the Philippines in 1991, and the eruption of Fourpeaked Mountain in Alaska in 2006.

Recharging 💬

Some volcanoes or volcano systems take several hundred thousand years to build up pressure between eruptions. This process is called **recharging** and primes the volcano for its next eruption.

Volcano Creation 😛

Different types of volcanoes are created by different forms of tectonic activity.

Volcano	Definition	Example
Subduction Zone	• How most volcanoes form • Occurs at convergent boundaries between oceanic and continental plates, and sometimes between two oceanic plates • The subducting plate is recycled into new magma, which rises through the overlying plate to create volcanoes.	Ring of Fire: A horseshoe-shaped zone of nearly uninterrupted volcanic features along the coast of South America. It was formed by convergent boundaries around the western, northern, and northeastern edges of the Pacific plate together with the eastern edge of the Nazca plate.
Rift Valley	• Occurs at divergent boundaries, usually between oceanic plates • New ocean floor is formed as magma fills in the gap between separating plates. • Thick magma is made of basaltic minerals and forms pillow lava when it interacts with cold ocean water. • Continental rift valleys also occur.	East African Rift: Forming where the Somali plate is breaking off from the African (or Nubian) plate in several places. A number of active and dormant volcanoes occur in the East African Rift, including Mount Kilimanjaro.
Hotspots	• Do not form at plate boundaries • Found in the middle of tectonic plates • Columns of unusually hot magma rise from deep in the mantle, partially melting and weakening the mantle in the lower lithosphere. • On oceanic plates, this leads to a chain of volcanic islands as the tectonic plate moves over a stable hotspot.	The Hawaiian Islands have been forming over a hotspot under the Pacific plate. The Yellowstone caldera (a system of volcanoes, or supervolcano) is an example of hotspot volcanism on a continental plate.

ASAP Environmental Science

Subduction Zone Volcanism 💬

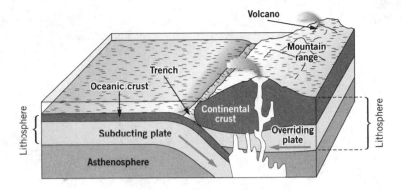

Hotspot Volcanism 💬

Chains of volcanic islands are created as the tectonic plate moves over a hotspot.

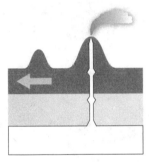 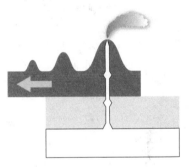

Types of Volcanoes ❗

Geologists and volcanologists classify volcanoes into four different types based on their shape, magnitude, structure, and type of eruption.

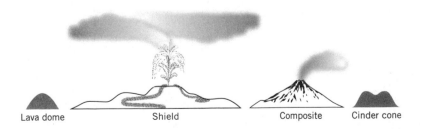

Lava dome   Shield   Composite   Cinder cone

Earth Systems and Resources

	Structure
Shield Volcanoes	• Tall • Broad base • Gentle slope
Composite Volcanoes	• Tall • Broad base • Steep slope
Cinder Cones	• Small and short • Steep slope • Symmetrical cone • Bowl-like crater at the summit
Lava Domes	• Small and short • Steep slope • Dome

Formation	Eruption	Examples
Over oceanic hot spots	• Usually mild • Slow lava flow	• Hualālai • Mauna Loa • Kohala
At subduction zones	• Violent eruptions • Eject lava, water, and gases as superheated ash and stones	• Mount Fuji • Mount St. Helens • Mount Rainier
Near other types of volcanoes	• Violent eruptions • Eject lava, water, and gases as superheated ash and stones	• Paricutin in Mexico • Lava Butte • Sunset Crater
Near other types of volcanoes	• Lava is too viscous to flow a great distance. • Lava hardens into a dome shape.	• Many within the crater of Mount St. Helens

 ## *Did You Know?*

Sometimes water can enter the vent of a shield volcano. This forms a pyroclastic flow—a fluidized mixture of hot ash and rock—and results in a very explosive eruption.

The Atmosphere ❗

The **atmosphere** is the envelope of gases held close to Earth by the force of gravity. The atmosphere is thin relative to Earth's size. The inner four layers of the atmosphere reach an altitude that's approximately 12.5% of Earth's radius!

The atmosphere consists of the following layers. Read this diagram from the bottom up!

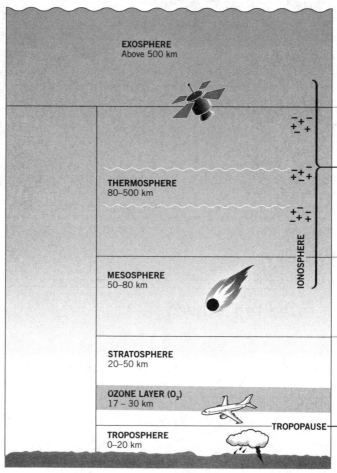

EXOSPHERE
Above 500 km

THERMOSPHERE
80–500 km

MESOSPHERE
50–80 km

IONOSPHERE

STRATOSPHERE
20–50 km

OZONE LAYER (O₃)
17 – 30 km

TROPOPAUSE

TROPOSPHERE
0–20 km

diagram not to scale

Did You Know?

Because of its density, the troposphere contains about 75–80% of Earth's atmosphere by mass, including greenhouse gases (see page 18).

EXOSPHERE
- Gases are thinnest here
- Leaks into outer space
- Upper boundary is a matter of dispute
- Some human-made satellites orbit here

THERMOSPHERE
- Gases are very thin
- Concentration of ionospheric bands results in high temperatures (e.g., 2000°C or 3632°F)
- Auroras (northern lights and southern lights) occur here
- Some human-made satellites orbit here

IONOSPHERE
- Not a distinct layer
- Includes ionized regions in the upper mesosphere, thermosphere, and lower exosphere
- Absorbs X-rays and ultraviolet radiation from Sun, leading to high concentration of ionic particles
- Ionically charged bands reflect radio waves, enabling long-distance radio communication

MESOSPHERE
- Air pressure extremely low
- Temperature decreases with altitude
- Atmosphere coldest at top of this layer: −90°C (−130°F)
- Where meteors usually burn up before striking Earth

STRATOSPHERE
- Similar in composition to troposphere but 1,000 times drier and less dense
- Includes the ozone layer
- Gases are not well-mixed
- Gradually warmer with altitude, because incoming heat from the Sun is trapped by the ozone layer
- Commercial passenger jets fly in the lower part of this layer

TROPOPAUSE
- Between the troposphere and the stratosphere
- Jet streams of air currents occur here

TROPOSPHERE
- Where weather takes place
- Contains 99% of the atmosphere's water vapor and clouds
- Usually well-mixed from bottom to top
- Gradually colder with altitude by 6.5°C/km or 3.5°F/1000 ft
- The air we breathe: 78% nitrogen + 21% oxygen
- Remaining 1% includes the greenhouse gases

Greenhouse Gases*		
Gas	Formula	% of the Troposphere
Water vapor	H_2O	0–4%
Carbon dioxide	CO_2	0.033%
Methane	CH_4	0.0002% (or 2 ppm)

Note that ppm stands for "parts per million." You'll learn more about this unit in Chapter 6.

⚠ Even though greenhouse gases make up only a small proportion of the atmosphere, they have a disproportionately large effect on Earth's environment.

Greenhouse Effect: Key Points ⚠

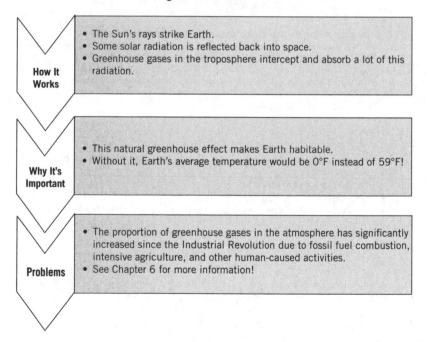

How It Works
- The Sun's rays strike Earth.
- Some solar radiation is reflected back into space.
- Greenhouse gases in the troposphere intercept and absorb a lot of this radiation.

Why It's Important
- This natural greenhouse effect makes Earth habitable.
- Without it, Earth's average temperature would be 0°F instead of 59°F!

Problems
- The proportion of greenhouse gases in the atmosphere has significantly increased since the Industrial Revolution due to fossil fuel combustion, intensive agriculture, and other human-caused activities.
- See Chapter 6 for more information!

Note that these are not the only greenhouse gases. We will discuss others in later chapters.

The greenhouse effect is an important concept for the AP Environmental Science Exam. We'll discuss the greenhouse effect in more detail later in this book.

Weather and Climate ⚠

Earth's atmosphere has physical features that change day to day as well as patterns that are consistent over many years.

Weather vs. Climate	
Weather	**Climate**
• Day-to-day properties • Wind speed and direction • Temperature • Amount of sunlight • Pressure • Humidity	• Patterns that are constant over many years (30 years or more) • Average temperature • Average precipitation amounts

Weather and climate are the result of Earth's rotation, as well as the Sun unequally warming the Earth and the gases above it. It is also affected by albedo, or reflectance.

Meteorologists are scientists who study weather and climate.

Albedo (Reflectance) 💬

Definition

Albedo, or **reflectance,** is the percentage of insolation (amount of solar radiation) reflected by a surface.

Albedo = 0

Absorb all radiation

Low Albedo

More solar radiation is absorbed; e.g., land and trees.

High Albedo

More solar radiation is reflected; e.g., snow and ice.

Albedo = 1

- Reflect all radiation

- Changes in albedo can alter the temperature.

Seasons 🔔

The motion of the Earth around the Sun, and Earth's axial tilt of 23.5 degrees, together create the seasons that we experience on Earth. This results in variation in the **insolation,** the amount of solar radiation that reaches a given area, of different parts of the Earth with the time of the year. When Earth is in the part of its orbit in which the Northern Hemisphere is tilted toward the Sun, the northern half of the planet receives more direct sunlight for longer periods of time each day than does the Southern Hemisphere. This means that when the Northern Hemisphere is experiencing summer, the Southern Hemisphere is experiencing winter.

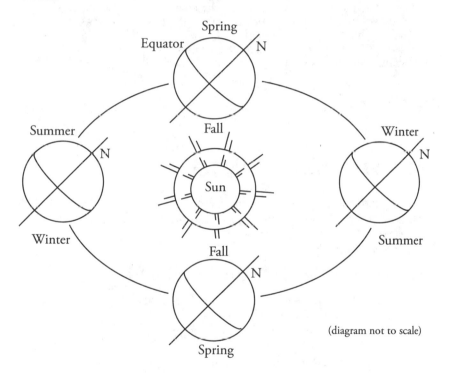

(diagram not to scale)

Seasons and Hemispheres ❗

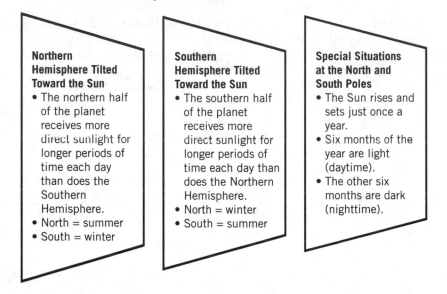

Northern Hemisphere Tilted Toward the Sun
- The northern half of the planet receives more direct sunlight for longer periods of time each day than does the Southern Hemisphere.
- North = summer
- South = winter

Southern Hemisphere Tilted Toward the Sun
- The southern half of the planet receives more direct sunlight for longer periods of time each day than does the Northern Hemisphere.
- North = winter
- South = summer

Special Situations at the North and South Poles
- The Sun rises and sets just once a year.
- Six months of the year are light (daytime).
- The other six months are dark (nighttime).

Atmospheric Circulation ❗

The motion of air around the globe is the result of solar heating, the rotation of Earth, the Coriolis effect (see the next page), and physical properties of air, water, and land.

3 Reasons Why Earth Is Unevenly Heated 〰

1. More of the Sun's rays strike Earth at the equator than strike the poles (per unit area).
2. The tilt of Earth's axis points regions toward or away from the Sun.
3. **The Coriolis effect:** See the diagram on the next page.

Earth's surface at the equator is moving faster than at the poles. This changes the motion of air into major prevailing winds, belts of air that distribute heat and moisture unevenly. Winds moving north from the equator in the Northern Hemisphere are deflected to the east, and winds moving south from the equator in the Southern Hemisphere are deflected to the west. This deflection pattern is known as the **Coriolis effect.**

The Coriolis Effect

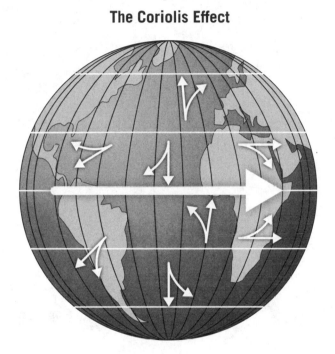

Convection Currents and Cells 😬

Solar energy warms Earth's surface, and heat is transferred to the atmosphere by radiation heating. The warmed gases expand, become less dense, and rise, creating vertical currents called **convection currents.** Warm currents can hold more moisture than surrounding air. As these large masses of warm moist air rise, cool air flows along Earth's surface into the area where the warm air was located. This flowing air or horizontal airflow is one way that surface winds are created.

Convection Currents

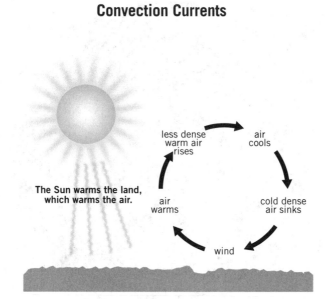

As warm moist air rises into the cooler atmosphere, it cools to the **dew point,** the temperature at which water vapor condenses into liquid water. This condensation creates clouds. If condensation continues and water drops get bigger, they can no longer be held up by the convection in Earth's atmosphere, and they fall as **precipitation** (which can be frozen or liquid). The cold, dry air is now denser than the surrounding air. This air mass then sinks to Earth's surface, where it is warmed and can gather more moisture, thus starting the convection cell rotation again.

Convection Cell

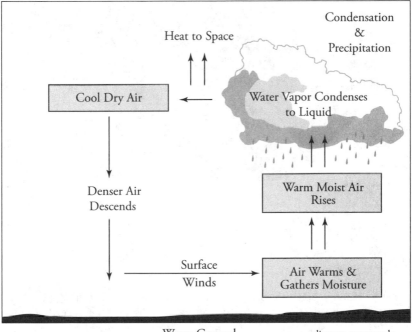

Global Convection Cells 💬

On a global scale, there are three sets of convection cells: **Hadley, Ferrel,** and **Polar cells**.

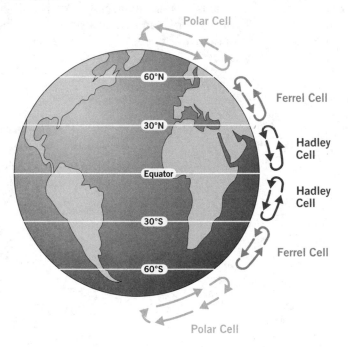

Hadley Cells 💬

A large Hadley cell starts its cycle over the equator. Warm, moist air evaporates and rises into the atmosphere. This leads to lots of precipitation (which is why there are so many equatorial rain forests). Also, cool, dry air descends about 30° north and south of the equator. This forms belts of deserts seen around those latitudes. Take a look at the following diagram.

ASAP Environmental Science

Hadley Cell

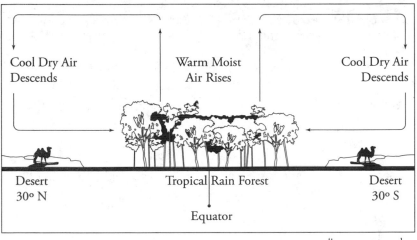

Cool Dry Air Descends

Warm Moist Air Rises

Cool Dry Air Descends

Desert
30° N

Tropical Rain Forest

Desert
30° S

Equator

diagram not to scale

What Is Wind? 🔊

Wind, part of Earth's circulatory system, is air that is moving due to unequal heating of Earth's atmosphere and flows from regions of high pressure to regions of low pressure. Wind moves heat, moisture, soil, and pollution around the planet.

Winds Around the World

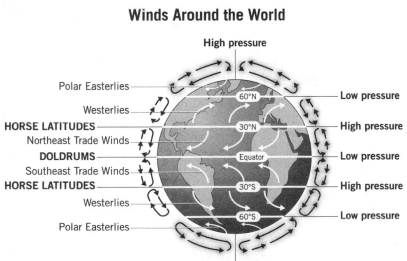

High pressure

Polar Easterlies

Westerlies

HORSE LATITUDES

Northeast Trade Winds

DOLDRUMS

Southeast Trade Winds

HORSE LATITUDES

Westerlies

Polar Easterlies

60°N

30°N

Equator

30°S

60°S

Low pressure

High pressure

Low pressure

High pressure

Low pressure

High pressure

System	Location	Due to	Characteristics	In the Northern Hemisphere	In the Southern Hemisphere
Trade Winds	Between the equator and 30° latitude	Hadley cell	• Steady and strong • Usually about 11–13 mph	NE to SW	SE to NW
Westerlies	30°–60° latitude	Ferrel cell	• Stronger in summer • Weaker in winter	SW to NE	NW to SE
Polar Easterlies	60° latitude and the poles	Polar cell	• Dry and cold winds • Often weak and irregular	NE to SW	SE to NW

 Did You Know?

Trade winds were named for their ability to quickly propel trading ships across the ocean as well as their historical importance to trading between continents.

Between the Winds

System	Location	Characteristics	Also Known as...
Doldrums	Between northeast and southeast trade winds, at the equator (5° on either side)	• Air is relatively still • Air is rising instead of blowing	Intertropical Convergence Zone (ITCZ)
Horse Latitudes	Between trade winds and westerlies (30° to 35° north and south of the equator)	• Air movement is less predictable • Weak wind due to subsiding dry air and high pressure	Subtropical high

---

 Did You Know?

Northeast and southeast trade winds converge in the ITCZ region, producing convectional storms and heavy precipitation.

Some people say that sailors gave the region of the subtropical high the name "horse latitudes" because ships relying on wind were unable to sail in these areas. Afraid of running out of food and water, sailors threw their horses (and other live cargo) overboard to save on food and water and to make the ship lighter and easier to move.

Occur in the upper troposphere

Fast-moving air currents

Have a large influence on local weather patterns

Found between global convection cells

Polar jet is usually around 60° latitude (between Polar and Ferrel cells)

Subtropical jet is usually around 30° latitude (between Ferrel and Hadley cells)

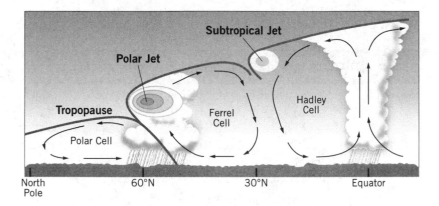

 Remember!

The tropopause is the interface between the troposphere and the stratosphere.

Atmosphere-Ocean Interactions ❗

Event	Definition	How It Forms	Where It Forms	Examples
Monsoon	• Seasonal winds • Usually accompanied by heavy rainfall • Occur because land heats up and cools down more quickly than water does	• Hot air rises from heated land • Low-pressure system is created • Rising air is quickly replaced by cooler moist air that blows in from over the ocean • Air rises and cools • Moisture is released in a steady seasonal rainfall	Coastal areas	Common in India and Southeast Asia between April and September
Rainshadow Effect	• Winds lose moisture on the windward side of a mountain range • Other side of the mountain receives relatively little precipitation	• Moist air from the ocean moves inland • Encounters an obstacle • Forced to climb in altitude • Air rises and cools • Moisture is released on the ocean or windward side of the barrier	Coastal areas with topographic barriers	West side of Olympic National Park in Washington State has a temperate rainforest (annual precipitation ~150 inches). The east side forests are much drier (annual precipitation ~15 inches). The Olympic mountain range divides the two.

Event	Definition	How It Forms	Where It Forms	Examples
Tropical Storms	• Localized • Very intense low-pressure wind system • Very strong winds	• Trade winds cause local disturbances when they blow over very warm ocean water • Air warms • Forms an intense, isolated, low-pressure system • Wind circles around this isolated low-pressure area	Over warm ocean waters near the equator	An everyday phenomenon that drives weather over much of the Earth
Hurricanes (Atlantic Ocean) Typhoons or Cyclones (Pacific Ocean)	• A giant and spiraling tropical storm with violent winds	• Rotating winds remove water from the ocean's surface • Heat is released as the water vapor condenses • Wind speed increases	Form over warm ocean waters near the equator	June through November in the Atlantic Anytime in the Pacific but mostly between July and September

Note that tropical storm winds circle counterclockwise in the Northern Hemisphere and clockwise in the Southern Hemisphere, once again because of the Coriolis effect.

How a Monsoon Forms ❗

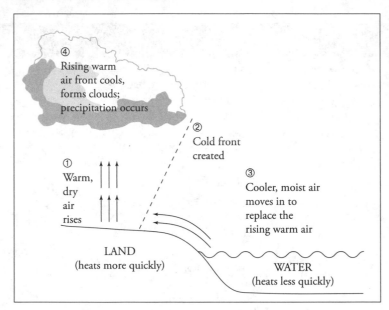

④ Rising warm air front cools, forms clouds; precipitation occurs

② Cold front created

① Warm, dry air rises

③ Cooler, moist air moves in to replace the rising warm air

LAND
(heats more quickly)

WATER
(heats less quickly)

How the Rainshadow Effect Works ❗

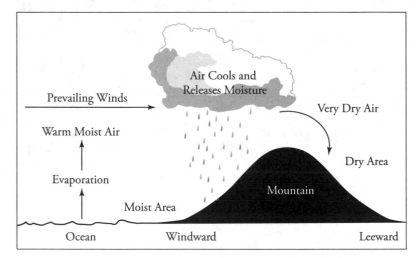

Prevailing Winds

Air Cools and Releases Moisture

Very Dry Air

Warm Moist Air

Dry Area

Evaporation

Mountain

Moist Area

Ocean   Windward   Leeward

Mountain ranges act as topographic barriers in the rainshadow effect.

El Niño and Southern Oscillation (ENSO) ❗

ENSO changes in atmospheric conditions lead to El Niños or La Niñas. These cycles involve powerful interactions between oceanic and atmospheric systems and exert a strong influence on the weather. **El Niño** is one extreme in a natural cycle, and **La Niña** is the opposite extreme.

Neutral Phase (Normal)

• Trade winds blow east to west across the Pacific Ocean
• Warm Pacific water moves from South American to Southeast Asia

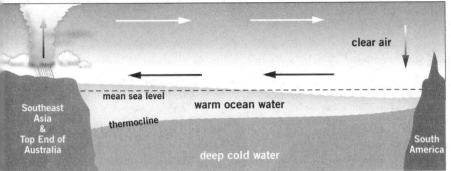

West Side of Pacific (Southeast Asia and top of Australia):
• Has a deeper top layer of ocean
• Warm surface water evaporates into the air
• More precipitation

East Side of Pacific (Off the coast of South America):
• Thermocline is nearer to the ocean surface
• Cold, nutrient-rich water has welled up into the upper parts of the ocean (upwelling)
• Coastal water relatively cool
• Air and weather are drier

The thermocline is a layer of the ocean that quickly decreases in temperature as you go down. The depth of the thermocline varies, but it is typically a few hundred meters below the ocean surface. The thermocline is a bit like an oceanic version of a temperature inversion (see Chapter 6) in the atmosphere.

❗ El Niño is a climate variation that takes place in the tropical Pacific, usually once every three to seven years.

El Niño Phase

- Reversal of the high and low pressure regions on either side of the tropical Pacific
- Trade winds lose their strength or reverse
- Warm water flows from Southeast Asia towards the coast of South America, along the equator

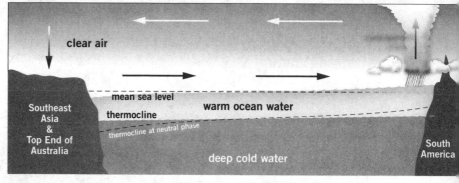

West Side of Pacific (Southeast Asia and top of Australia):

- Thermocline is nearer to the ocean surface
- Cold, nutrient-rich water has welled up into the upper parts of the ocean (upwelling)
- Air and weather are drier than normal

East Side of Pacific (Off the coast of South America):

- Upwelling slows or stops
- Has a deeper top layer of ocean
- Water off the coast becomes warmer and contains fewer nutrients
- Some offshore fish populations decline (negative economic effects)
- Warm surface water evaporates into the air
- Higher-than-average precipitation

During El Niño years, the northern United States and Canada experience warmer winters and a less intense hurricane season. The effect of climate change on these cycles is not yet understood, though some scientists think El Niño will become more common in the future.

La Niña Phase

- Reverse of El Niño
- Due to the Coriolis effect
- Air moves toward the equator to replace rising hot air
- Moving air deflects to the west and helps move surface water
- Enhanced upwelling

West Side of Pacific: Surface waters are warmer than usual
East Side of Pacific: Surface waters are colder than usual

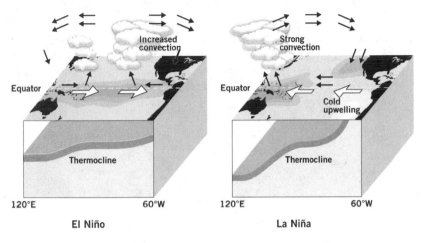

El Niño   La Niña

💬 El Niño and La Niña events usually develop during the southern winter, intensify during the southern spring, peak around December, and return to neutral conditions during the southern autumn. In other words, the typical life cycle of an event is about 12 months.

Global Water Resources and Use ❗

The **hydrosphere** includes Earth's oceans and freshwater bodies. Water covers about 75% of planet Earth.

Freshwater ❗

Freshwater contains only minimal quantities of dissolved salts, especially sodium chloride. All freshwater ultimately comes from precipitation of atmospheric water vapor, which reaches inland lakes, rivers, and groundwater bodies directly, or after melting of snow or ice.

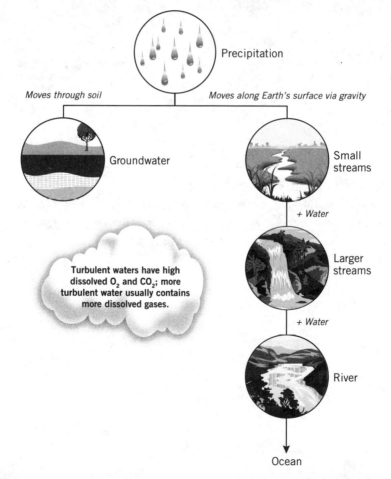

Precipitation

Moves through soil

Moves along Earth's surface via gravity

Groundwater

Small streams

+ Water

Turbulent waters have high dissolved O_2 and CO_2; more turbulent water usually contains more dissolved gases.

Larger streams

+ Water

River

Ocean

A **watershed** is an area of land that collects rainwater and drains it into a particular stream or river.

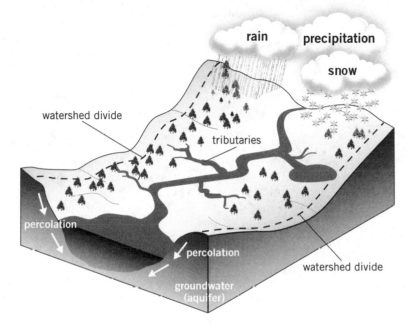

Freshwater that travels on land is largely responsible for shaping Earth's surface

Travels from the highlands to the ocean

Doesn't move in straight lines because of obstructions on land

Follows the lowest topographical path

Follows the path of least resistance

Water slows down around bends and drops some of its sedimentary load

Erosion

• Movement of water etches channels into rocks

• Moving water carries eroded material farther downstream

Examples of Freshwater Biomes 💬

Deltas	Estuaries	Wetlands
• Landforms that form at the juncture of a river and ocean • Made of deposited sediment • Rivers drop most of their sedimentary load as they meet the ocean because their velocity decreases	• Sites where the "arm" of the sea extends inland to meet the mouth of a river • Often rich with many different types of plant and animal species • Water usually has a high concentration of nutrients and sediments • Shallow and warm water • Examples: salt water marshes, mangrove forests, inlets, bays, and river mouths	• Ecologically diverse ecosystems • Found along the shores of fresh bodies of water • Examples: marshes, swamps, bogs, prairie potholes, and flood plains

Vertical Stratification in Freshwater Biomes 💬

In all natural bodies of water, there are layers of water that vary significantly in temperature, oxygen content, and nutrient levels. These layers are affected by seasonal changes and other disturbances.

Epilimnion
- Warm, typically oxygen-rich
- Recurrent surface scum
- Floating weed masses

Metalimnion (Thermocline)
- Zone of decreasing water temperature
- Zone of decreasing dissolved oxygen

Hypolimnion
- Zone completely void of oxygen
- Cold water temperatures
- Accumulating organic muck
- Weeds prevalent
- Unusable by fish and all aerobic organisms

Freshwater Zones 💬

These layers of water are also often delineated based on the types of organisms that can live in them.

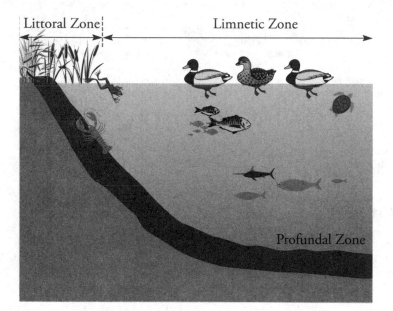

___

Some bodies of freshwater also have a benthic zone. This is the deepest layer, characterized by very low temperatures and low oxygen levels.

Zone	Location	Oxygen Content	Sunlight	Temperature	Organisms
Littoral	Shallow water at the shoreline	High	High	Warm	• Phytoplankton • Zooplankton • Small crustaceans • Many freshwater fish
Limnetic	Surface of open water	High	High	Moderate	• Phytoplankton • Zooplankton • Swimming insects and fish
Profundal	Deep water	Low	None (aphotic)	Coolest	• Detritivores • Leeches and other annelid worms • Some species of insect larvae • A few types of crabs and mollusks

Did You Know?

Phytoplankton are autotrophs: they perform photosynthesis and release oxygen into the water.

Saltwater ❗

Most of the water on Earth's surface is saltwater, or seawater. However, seawater is not uniformly saline throughout the world. The planet's freshest seawater is in the Gulf of Finland, part of the Baltic Sea. The most saline open sea is the Red Sea, where high temperatures and confined circulation result in high rates of surface evaporation.

Examples of Saltwater Biomes 💬

Barrier Islands	Coral Reef
• Landforms that lie off coastal shores • Spits of land • Created by buildup of deposited sediments • Boundaries constantly shifting as water moves around them • Function to buffer the shoreline behind them	• A type of barrier island • Formed by a community of living things • Dominant organisms: cnidarians, which secrete a hard, calciferous shell • Cnidarians associate with colorful algae, which provide them with food • Home to a large diversity of species • Extremely delicate • Vulnerable to physical stresses, changes in light intensity, changes in water temperature, and ocean depth and pH

Saltwater Zones 💬

Like freshwater bodies, oceans are divided into zones based on changes in light and temperature.

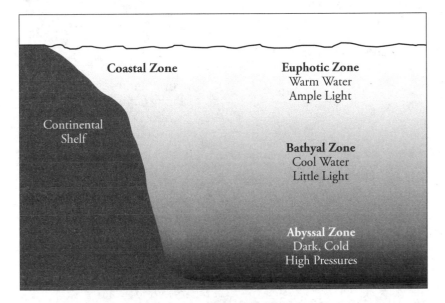

Zone	Location	Oxygen Content	Sunlight	Temperature
Coastal	Ocean water closest to land, between the shore and the end of the continental shelf	High	High	Warmest
Euphotic	Surface of open water	High	High	Warm
Bathyal	Deep water	Low	None (aphotic)	Cool
Abyssal	Deepest region of the ocean	Very low	None (aphotic)	Extremely cold

---

The abyssal zone has very high levels of nutrients because decaying plant and animal matter sinks down from the zones above.

Ocean Circulation and Currents ❗

Ocean currents play a major role in modifying conditions and climates around Earth.

Ocean water moves due to wind, differences in salinity (saltiness), and Earth's rotation.

The Sun warms water near the equator.

In the Northern Hemisphere, the Gulf Stream carries sun-warmed water north, along the East Coast of the U.S.

Warm water displaces colder, denser water in polar regions.

Cool water moves south and is re-warmed by the equatorial sun.

Ocean Conveyor Belt

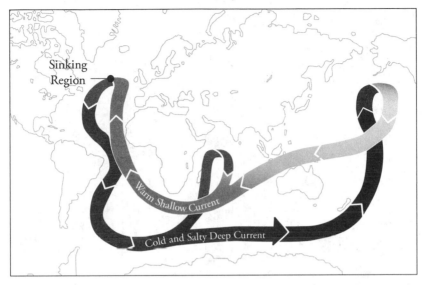

The oceans have a known pattern of flow called the ocean conveyor belt. The warm Gulf Stream air cools as it hits northern latitudes, sinks, and is pushed toward the south pole. Eventually it makes its way back to the surface where it can become part of the Gulf Stream again.

Water Use 💬

Humans use water for agriculture and crop irrigation (73%), industry (21%), and domestic use (6%).

Where Does Our Water Come From?

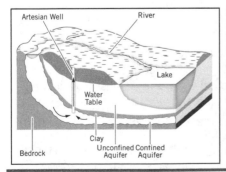

Groundwater: Water present beneath Earth's surface, in soil pore spaces and in fractures of rock formations

Aquifer: A type of groundwater that yields a usable quantity of water

Unconfined Aquifer: Water seeps into the aquifer from the ground surface directly above

Confined Aquifer: Surrounded by an impermeable dirt/rock layer, which prevents water seeping in from the ground above

Water Use in the United States

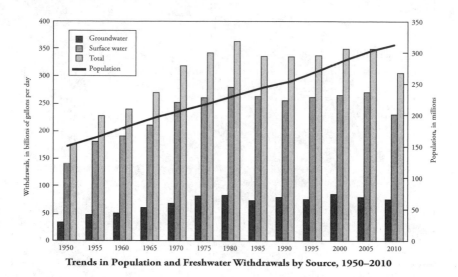

Trends in Population and Freshwater Withdrawals by Source, 1950–2010

Water Problems and Issues ❗

> ❗ Since the 1950s, global water use has tripled, mostly due to population growth and improvements in the global standard of living.

- **Water risk:** The possibility of an area experiencing a water-related challenge such as *water scarcity, water stress,* flooding, or drought
- **Water stress:** Occurs when demand for water exceeds the available amount during a certain period, or when poor quality restricts water use. Water-stressed countries have a renewable annual water supply of about 1,000–2,000 m³ per person.
- **Water scarcity:** A lack of sufficient available water resources to meet demand for water in a region. It affects every continent and around 2.8 billion people around the world at least one month out of every year. Water-scarce countries have a renewable annual water supply of less than 1,000 m³ per person.
  - **Examples:** Algeria, Egypt, Libya, Kenya, Rwanda, Tunisia, Israel, Jordan, Kuwait, Saudi Arabia, Syria, Belgium, Hungary, the Netherlands, Singapore, Barbados, Malta, Lebanon, Morocco, Niger, Somalia, South Africa, and Sudan

ASAP Environmental Science

💬 The United States as a whole is not considered water-scarce, but certain regions are considered water-stressed. Additionally, the U.S. is using water more quickly than it can be replenished, so water scarcity is possible in the near future.

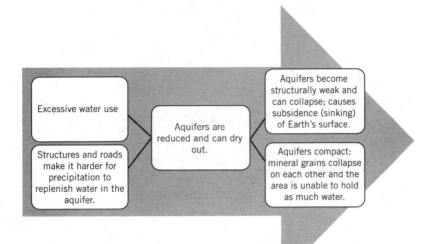

Water Conservation 🔔

Water conservation includes policies, strategies, and activities to sustainably manage the natural resource of fresh water, to protect the hydrosphere, and to meet the current and future human demand for water. Water conservation efforts and greater efficiencies in using water have had a positive effect over the last 35 years.

Many water-scarce countries are developing countries that have rapidly increasing populations. This means their water scarcity problems will likely continue to grow over time.

Human Water Rights ❗

Challenges

It's difficult for politicians and lawmakers to put restrictions on water use because many people think it should be free.

Riparian Rights

Water rights are given to those who have legal rights to use that area.

Prior Appropriation

Water rights are given to those who have historically used water in a certain area.

Looking Forward

As water becomes scarcer globally, it will be important for countries to think of ways to regulate the use of water.

Desalinating ocean water has been proposed as a solution to our current global water crises. However, distillation or reverse osmosis uses a large amount of energy and is not economically viable.

Soil and Soil Dynamics ❗

The **pedosphere** is more commonly known as soil. Soil plays a crucial role in the lives of the plants, animals, and other organisms that live in the biosphere. The pedosphere exists at the interface of the other four systems or spheres (lithosphere, hydrosphere, atmosphere, and biosphere).

What Is Soil? ❗

Soil is a complex, ancient material teeming with living organisms. Some soil is hundreds of years old! It contains a mixture of:

- Finely broken-down or weathered rock (45%)
- Living and dead organic matter (~5%), such as protozoa, bacteria, algae, fungi, and animals
- Air (~25%)
- Water (~25%)

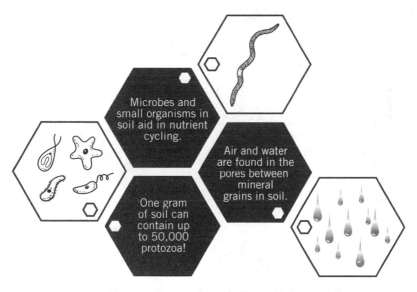

Microbes and small organisms in soil aid in nutrient cycling.

Air and water are found in the pores between mineral grains in soil.

One gram of soil can contain up to 50,000 protozoa!

The Chemistry of Soil: pH ❗

The pH of most substances ranges from 0–14.

pH is a measure of the concentration of hydrogen ions.

Most soils have a pH between 4 and 8.

Soil can be **acidic** (pH < 7), **neutral** (pH = 7), or **alkaline/basic** (pH > 7).

Soil pH affects nutrient solubility and thus nutrient availability to plants.

❗ When the pH of soil becomes too acidic, ions of heavy metals such as mercury (Hg) or aluminum (Al) can leach into the groundwater. These ions can travel to streams and rivers and harm both plants and aquatic life.

Why Is Soil Important?

Links the abiotic (nonliving) and biotic (living) components of the world

Plays an active role in nutrient cycling

Supports plant life by supplying a physical foundation for growth, water, and nutrients

Nonrenewable resource

 It takes 500 to 1,000 years to form a single inch of soil, and at least 3,000 years to form enough fertile soil to support crop growth.

Rocks !

The oldest rocks on Earth are 3.8 billion years old, while others are only a few million years old.

Rock Cycle !

Time, pressure, and the Earth's heat interact to create three basic types of rocks.

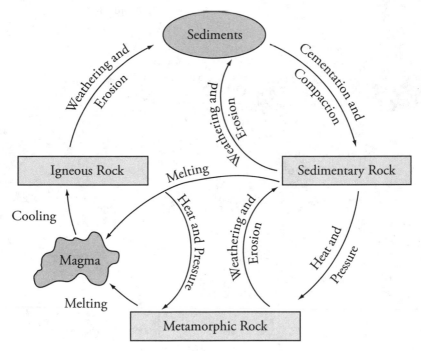

Types of Rocks 💬

Rock	How It Is Made	Example
Igneous (Magmatic) Rock	• Rock is melted by heat and pressure below the crust. • Magma comes to the surface of the Earth, emerges as lava, and re-solidifies into igneous rock.	• Basalt • Granite
Sedimentary Rock	• Formed as sediment (eroded rocks and the remains of plants and animals) builds up and is compressed • Forms under water as sediments or dissolved minerals deposit on a stream bed or ocean floor. They are compressed as more material is deposited and then cemented together.	• Limestone • Shale • Coal • Chert
Metamorphic Rock	• Formed by transforming an existing rock via a large amount of pressure and heat • Can happen as sedimentary rocks sink deeper into the earth and are heated by the high temperatures found in the Earth's mantle	• Slate • Marble

- **Molten rock** refers to rock that has been melted.
- Molten rock above ground is called **lava.**
- Molten rock underground is called **magma.**

Rock Weathering 🕛

Soil is a combination of organic material and rock that has been broken down by different types of weathering: physical, chemical, or biological.

Type of Weathering	How It Works	Examples
Physical (Mechanical)	• Rock is weakened and worn down by physical forces, usually wind or water. • No change in the chemistry of the rock • Dominates in cold or dry environments	• Water trapped in small surface cracks of rocks expands when frozen and enlarges the cracks. • Extreme heat, especially from fire, causes thin layers of rock to flake away at the surface. • Abrasion of rock surfaces by water, ice, and wind
Chemical	• Rock is subjected to chemical alteration through reactions with water, oxygen, or dissolved minerals. • Breakdown and reorganization of rock minerals • Dominates in warm or moist environments	• Rust forms when iron and other metallic elements come in contact with water.
Biological	• Weathering that takes place as the result of the activities of living organisms • Organisms can weather rock by physical or chemical means.	• Tree roots enlarge cracks as they grow (physical). • Plant roots or lichens growing on rocks release organic acids, which dissolve minerals (chemical).

 Did You Know?

Many wetland soils aren't made from a base of weathered rocks. Instead, they originate from organic matter that decomposes slowly in a waterlogged environment and becomes thick enough to serve as a medium for plant growth.

Types of Soil ❗

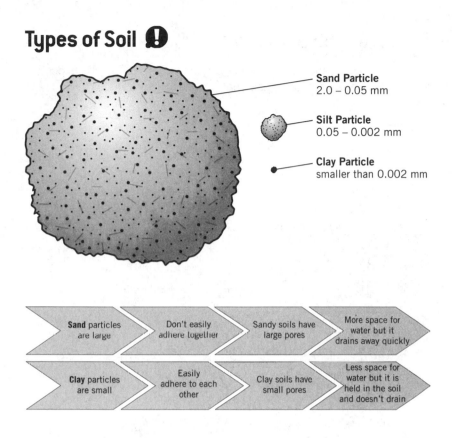

Sand Particle
2.0 – 0.05 mm

Silt Particle
0.05 – 0.002 mm

Clay Particle
smaller than 0.002 mm

| **Sand** particles are large | Don't easily adhere together | Sandy soils have large pores | More space for water but it drains away quickly |

| **Clay** particles are small | Easily adhere to each other | Clay soils have small pores | Less space for water but it is held in the soil and doesn't drain |

Loam is soil containing a blend of sand, silt, and clay, in varying proportions. Loamy soil is best for plant growth.

Composition: Soil Layers ❗

Soil comprises distinct layers known as horizons, which vary considerably in content. Not every type of soil contains all the horizons.

Soil Horizons ❗

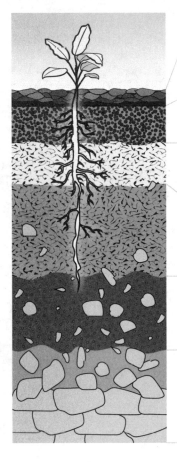

O HORIZON — *Humus/Organic*
Organic matter at various stages of decomposition. Includes animal waste, leaves and other plant tissues (such as dead roots and the decomposing bodies of organisms. The stable residue left after most organic matter has decomposed is a dark, crumbly material called humus.

A HORIZON — *Topsoil*
Topmost mineral horizon; this is the most intensively weathered soil layer. Its dark color is due to accumulation of organic matter from the O horizon. This is the zone of leaching.

E HORIZON — *Zone of Eluviation*
Light in color and coarse in texture; no organic matter has traveled down from the A horizon, while clays and minerals like iron and aluminum oxides have been washed out. Usually found in soils developed under forest.

B HORIZON — *Zone of Illuvation*
Sometimes called the subsoil, this is where organic matter, clay and minerals washed out of the upper horizons accumulate.

C HORIZON — *Parent Material*
Made up of unconsolidated material, loose enough to be dug up with a shovel. Weathering at this depth is minimal so the soil retains identifiable structural features of the parent material (the rock from which the A and B horizons formed). This horizon has much less biological activity than the horizons above it.

R HORIZON — *Bedrock*
This layer of consolidated, unweathered rock is not, strictly speaking, part of soil. If the material from which horizons A through C formed was *not* transported from elsewhere, the R horizon is the same as the parent material.

Eluviation is the movement of dissolved material from higher soil layers to lower soil layers due to the downward movement of water, which is caused by gravity. Illuviation is the deposition of minerals, humus, and other materials in a soil horizon.

ASAP Environmental Science

Soil Development 🌀

The development of soil horizons is accomplished by processes that work on soil minerals and particles, organic matter, and soil chemistry by altering and moving these components through the soil profile. Soil development is influenced by six soil-forming factors.

Soil-Forming Factors (CI.O.R.P.T.H.)	
Climate	• Involves differences in temperature and precipitation across the globe • Both heat and water facilitate chemical and biochemical reactions. • Seasonal fluctuations contribute to rock weathering. • Determines what organisms grow in a particular location
Organisms (Biological Activity)	• Different local conditions support different organisms, which influence soil. • Microorganisms decompose organic matter and transformation minerals. • Animals move soil, consume vegetation, and add nutrients through waste and decomposing bodies. • Plants perform physical weathering through root growth, take up soil nutrients and water, alter soil chemistry, and add nutrients when they die and decompose.
Relief (Topography)	• Topographical relief affects water flow and depth to the water table. • Affects erosion: which locations are likely to lose surface material, and which are likely to accumulate eroded material • Leads to differences in how much sun different locations receive • Influences what and how organisms grow in a particular location

Parent Material	• Starting point for soil development
	• Mineral properties, hardness, and topography affect how it is weathered into soil.
	• Examples:
	  ○ Parent material rich in quartz (such as granite and sandstone) weathers into sandy soil.
	  ○ Shale weathers into soil richer in silt and clay.

Time	• More time means more change! • Hard parent material weathers more slowly, and softer material weathers more quickly. • A flat, stable topographic position develops more quickly than slopes and depressions where material is lost and gained.
Human Influence	• Use of fertilizer, pollution, and acid rain alter soil chemistry • Construction activities like digging and plowing mix soil and blur distinctions between horizons. • Traffic and machinery compacts soil. • Removing vegetation leads to more erosion. • Irrigation and depletion of groundwater leads to salinization (high salt content).

Soil Problems 🔱

In order to grow food, we must have enough **arable** (suitable for plant growth) soil to meet our agricultural needs. **Soil fertility** refers to soil's ability to provide essential nutrients—such as nitrogen, potassium, and phosphorus—to plants. **Humus** is also extremely important because it is rich in organic matter.

Erosion 🔱

Erosion is the removal of soil, rock, or dissolved material from one location on the Earth's crust and transportation to another location. Soil without any plants growing in it is more susceptible to erosion than soil that is covered by organic materials. Because of the constant movement of water and wind on Earth's surface, erosion is a continual and normal process.

Human activities such as overcultivation of agricultural fields, over-grazing, urbanization, and deforestation have significantly increased the levels of erosion in the upper layers of soil. Eroded topsoil usually ends up in bodies of water, posing a problem for both farmers, who need healthy soil for planting, and people who rely on bodies of water to be uncontaminated with soil runoff.

Monoculture vs. Polyculture ❗

Monoculture	Polyculture
• Planting just one type of crop in a large area • Common in modern agriculture • Leaches soil of specific nutrients • Decreases crop genetic diversity, making crops more susceptible to pests and diseases	• Planting many types of crops in a large area • Increases sustainability • Limits nutrient depletion in soil • Increases genetic diversity • Crop rotation another good solution

Other Problems with Modern Agriculture ❗

- Damaged soil due to large machinery
- Repeated plowing breaks down soil aggregates; this leaves *plow pan,* or *hard pan,* which is hard, unfertile soil.
- Large energy consumption, including burning fossil fuels
- Pollution in the form of pesticides and fertilizers

Green Revolution ❗

What Is It?	Cons	Solutions
• A huge increase in worldwide agricultural productivity over the past 50 years • Due to the mechanization of farming that resulted from the Industrial Revolution	• Many detrimental effects on the environment • Use of chemical pesticides lead to new pesticide-resistant insects • Over-irrigation has caused salinization of soil and land degradation	• Genetically modified plants decrease the need for pesticide use. • Drip irrigation adds only the water required and delivers water directly to plants' roots.

Soil Conservation ❗

Much of what we know about soil conservation was established relatively recently. Several best management practices have been developed to promote soil conservation. These practices return organic matter to the soil, slow down the effects of wind, and reduce the amount of damage done to the soil by tillage (plowing). These practices include the following:

- Using animal waste (manure), compost, and the residue of plants to increase the amount of organic matter in the soil
- Organic agriculture, a method of farming that uses compost, manure, crop rotation, and non-chemical methods to manage soil fertility and pest control
- Changing tillage (plowing) practices to reduce soil breakup and erosion, including contour plowing and strip planting
- Using trees and other wind barriers to reduce wind force

Soil Laws 💬

Date	Name	What It Did
1935	Soil Conservation Act	• Led to the creation of the Soil Conservation Service • Created in response to the Dust Bowl of the 1930s • Created to conserve soil and restore the nation's ecological balance • Led by soil conservation pioneer Hugh Hammond Bennett
1977	Soil and Water Conservation Act	• Established soil and water conservation programs to aid landowners and users • Set up conditions to continue evaluating the condition of U.S. soil, water, and related resources
1985	Food Security Act	• Discouraged the conversion of wetlands to non-wetlands • Nicknamed the Swampbuster • Led to future federal legislation (in 1990) that denied federal farm supplements to those who converted wetlands to agriculture, and provided a restoration of benefits to those who converted lands to wetlands

Did You Know?

Chapter 1 Summary ❗

The purpose of this chapter was to review the five interconnected spheres that the area near the surface of the Earth is divided into: lithosphere and pedosphere (land), hydrosphere (water), atmosphere (air), and biosphere (living things). The biosphere will be discussed more in the next chapter.

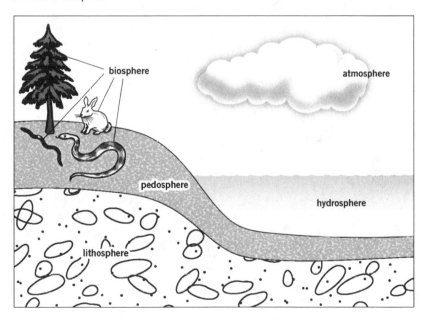

CHAPTER 2

The Living World

Living things are organized into specialized groups: the biosphere contains ecosystems, which contain communities, which contain populations of unique species. There are both terrestrial and aquatic ecosystems. Species in populations have distinct roles and interact with each other and the environment. Energy and nutrients continuously cycle through organisms and the environment. Both organisms and ecosystems change over time, in response to continually changing conditions.

Ecosystem Structure ❗

Abiotic vs. Biotic Components 〰

The Earth is made of both abiotic and biotic components.

Abiotic
Nonliving components of Earth

Examples: Atmosphere, hydrosphere, and lithosphere

Biotic
Living components of Earth

Examples: Animals, plants, fungi, protists, and bacteria, which form the biosphere

Organisms Interact with the Environment ❗

Population

A group of organisms of the same species

Community

- Populations of different species that occupy the same geographic area
- Every species occupies a habitat and has an ecological niche

Habitat

The area or environment where an organism lives, or where an ecological community occurs

Ecological Niche

- The role and position a species has in its environment
- Includes how a species uses biotic and abiotic resources in its environment, where the species lives, and what it eats

 Did You Know?

Some species interact quite a bit with other members of their population and others do not. For example, ants live in colonies and wildebeests live in herds, while bears don't tend to live in large groups.

Organisms Interact with Each Other ❗

There are three types of relationships between species: **competition, predation,** and **symbiotic** relationships.

Competition ❗

Competition occurs when two individuals—of the same species or of different species—are competing for resources in the environment, such as food, air, shelter, sunlight, or a mate. The "most fit" competitor will eventually win (and obtain the resource), and the others will be eliminated.

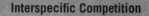

Intraspecific Competition

Between two individuals of the same species

Interspecific Competition

Between two individuals that are different species

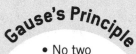

Competitive Exclusion

When two different species in a region compete and the better adapted species wins

Gause's Principle

• No two species can occupy the same niche at the same time

• The less fit species will relocate, die out, or occupy a smaller niche

Resource Partitioning

• A way to avoid direct competition

• Different species use slightly different parts of a habitat, but rely on the same resource

Competition can affect the niche an organism occupies.

Fundamental Niche	• The niche an organism inhabits in the absence of competition
Realized Niche	• The niche an organism inhabits in the presence of competition • Smaller than the fundamental niche

Here is an example of **resource partitioning:** five different species of warblers can all live in the same pine tree. They can coexist because each species feeds in a different part of the tree.

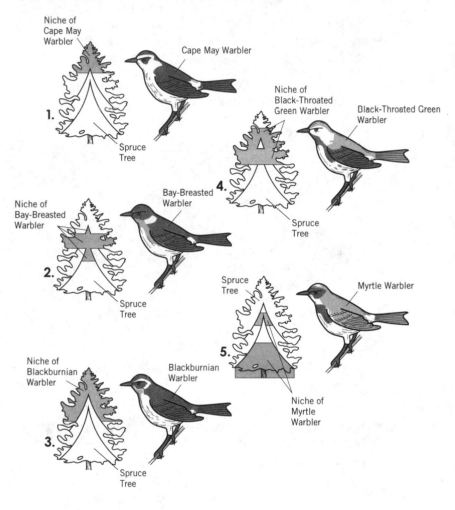

Did You Know?

Competition occurs between animals as well as plants. Plants compete for sunlight and for ground space, and some even produce chemicals to inhibit the growth of other plants.

Predation ❗

Predation occurs when one species feeds on another, and thus drives changes in population size.

Examples of Predation

- A cat and mouse are a classic example of a predator/prey relationship.
- Polar bears prey on seals in northern climates.
- Lions prey on wildebeests in the African savannah.
- Great white sharks prey on elephant seals in the marine biome.
- Osprey prey on fish in freshwater biomes.

 Did You Know?

It is easy to think of examples of animal predators, but there is more to ecology than animals! Herbivores prey on plants, and zooplankton prey on phytoplankton.

Symbiotic Relationships ❗

Symbiotic relationships are close, prolonged associations between two or more different organisms of different species. There are three different types of symbiosis.

Symbiotic Relationship	One Organism	The Other Organism	Example 〰
Mutualism	☺	☺	Pollinators benefit from feeding on plant nectar, and plants benefit from pollinators moving their pollen from plant to plant.
Commensalism	☺	😐	Barnacles are sedentary crustaceans that sweep the surrounding water for small, free-floating organisms. They often grow attached to scallop shells, which has almost no effect on the scallop.
Parasitism	☺	☹	Mistletoe is a flowering plant that grows attached to, and within, the vascular system of a tree or shrub. This helps the mistletoe but harms the tree.

 Ask Yourself...

Oxpeckers are a type of bird found in Africa. They eat the ticks and parasites off of rhinos. What kind of symbiotic relationship is this an example of?

Parasites and predators both live by feeding on another organism. However, predators kill the other organism while many parasites do not.

Types of Ecosystems ❗

Before we review types of ecosystems, let's look at some terms you should be comfortable using.

Blending

Remember that biomes blend into each other. There are no distinct boundaries.

Ecotones

A transitional area where two biomes meet

Ecozones or Ecoregions

Small regions within ecosystems that have similar physical features

Edge Effects

Ecotones have a great amount of species diversity and biological density. Some species only live on the edge of certain habitats.

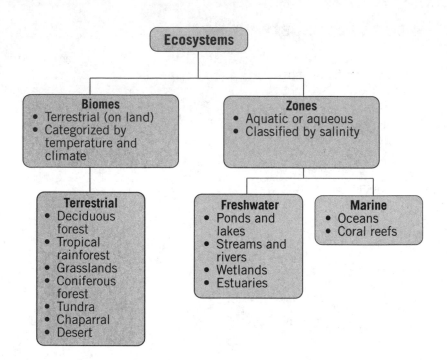

Ecosystems

Biomes
- Terrestrial (on land)
- Categorized by temperature and climate

Zones
- Aquatic or aqueous
- Classified by salinity

Terrestrial
- Deciduous forest
- Tropical rainforest
- Grasslands
- Coniferous forest
- Tundra
- Chaparral
- Desert

Freshwater
- Ponds and lakes
- Streams and rivers
- Wetlands
- Estuaries

Marine
- Oceans
- Coral reefs

 Remember!

The two most important factors in describing climate are average temperature and average precipitation amounts.

Types of Ecosystems ❗			
Biome	**Annual Rainfall, Soil Type**	**Major Vegetation**	**World Location**
Deciduous forest (temperate and tropical)	75–250 cm, rich soil with high organic content	Hardwood trees	North America, Europe, Australia, and Eastern Asia
Tropical rainforest	200–400 cm, poor quality soil	Tall trees with few lower limbs, vines, epiphytes, plants adapted to low light intensity	South America, West Africa, and Southeast Asia
Grasslands	10–60 cm, rich soil	Sod-forming grasses	North American plains and prairies; Russian steppes; South African velds; Argentinian pampas
Coniferous forest (Taiga)	20–60 cm— mostly in summer, soil is acidic due to vegetation	Coniferous trees	Northern North America, northern Eurasia
Tundra	Less than 25 cm, soil is permafrost	Herbaceous plants	The northern latitudes of North America, Europe, and Russia
Chaparral (scrub forest)	50–75 cm— mostly in winter, soil is shallow and infertile	Small trees with large, hard leaves, spiny shrubs	Western North America, the Mediterranean region
Deserts (cold and hot)	Less than 25 cm, soil has a coarse texture (i.e., sandy)	Cactus, other low-water adapted plants	30 degrees north and south of the equator

What Limits How a Species Can Grow? 💬

Two factors that affect how a population of a certain species grows are known as the Law of Tolerance and the Law of the Minimum.

Law of Tolerance 💬

The **Law of Tolerance** is the degree to which living organisms are capable of tolerating changes in their environment.

Living organisms and individual organisms exhibit a range of tolerance. For example, the sugar maple cannot tolerate average monthly high temperatures above 24–27°C or winter temperatures below −18°C.

Each organism has a certain minimum, maximum, and optimum combination of environmental factors that determine success.

Tolerance ranges are not always fixed. They can change as seasons, environmental conditions, or life stage of the organism changes. For example, the eggs and larvae of blue crabs require higher salinity than adult blue crabs.

This concept is the basis for **natural selection** (more on this later), which drives evolution. Individuals in a species that are capable of surviving change (due to their particular genetic makeup) will thrive. Those that are incapable will die off and not contribute their genes to the future population.

Numbers of Individuals in a Population That Can Tolerate Variations in an Environmental Factor

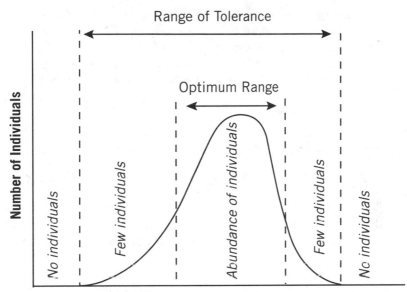

Range of Tolerance

Optimum Range

Number of Individuals

No individuals

Few individuals

Abundance of individuals

Few individuals

No individuals

Magnitude of an Environmental Factor

Law of the Minimum 💬

Minimum

- Growth is dictated by the scarcest resource (limiting factor), not by the total resources available.

- For example, if you increase the amount of plentiful nutrient, crop growth will **not** increase. If you increase the amount of a limiting nutrient, crop growth **will** increase.

- This is similar to how the amount of water a barrel can hold is limited by the shortest stave*.

😎 Vocab time! A stave is a slat in the wall of a barrel.

Energy Flow ❗

Life depends on a large number of costly reactions that occur within each cell. Many of these reactions are powered by ATP, a high energy molecule found in cells. Organisms make ATP from carbohydrates and other biomolecules in a complex series of reactions called **cell respiration.** Instead of eating a stack of sugary pancakes like we can, plants harvest solar energy and use it to make the starting carbohydrates for cell respiration.

Bioenergetics

Bioenergetics is the study of how energy flows through living organisms. Chemical bonds are the source of all energy for life.

- All energy on the Earth comes from the Sun.
- Photosynthetic organisms use solar energy to turn CO_2 and water into carbohydrates.
  - This also releases O_2.
- Biological macromolecules, especially carbohydrates and fats, store energy in their chemical bonds.
- Cells use respiration to transfer energy from the chemical bonds of macromolecules to the chemical bonds of smaller molecules such as ATP, GTP, NADH, NADPH, and $FADH_2$.
  - This releases CO_2 and H_2O.
  - These small, high-energy molecules power all the reactions and processes the cell needs to perform.

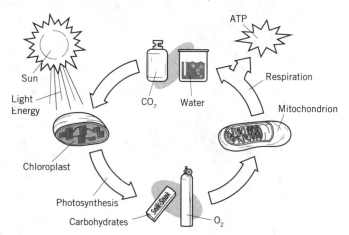

Photosynthesis ❗

Plants and algae are producers and perform photosynthesis: they convert solar (light) energy into chemical energy.

In plants, most photosynthesis occurs in leaves.

Algae includes some bacteria (blue-green algae or cyanobacteria) and plant-like protists (other algae).

Photosynthesis occurs in two steps: **light-dependent reactions** and **light-independent** (or dark) **reactions.** The light-dependent reactions produce the chemical energy needed by the light-independent reactions to produce organic molecules.

Photosynthetic Reactions		Purpose	Reactant(s)	Product(s)
	Light	Convert light energy to chemical energy (ATP and NADPH)	Sunlight H_2O	O_2 ATP NADPH
	Dark	Use ATP and NADPH to build organic molecules from CO_2	CO_2 ATP NADPH	Carbohydrates

Cell Respiration ❗

All living organisms require a source of energy, and all living organisms use the same metabolic pathways to power life.

All living organisms perform cellular respiration.

Cell respiration is a series of redox reactions. Glucose is oxidized to CO_2, and this powers the synthesis of ATP, an energy storing molecule.

Carbon Flow ❗

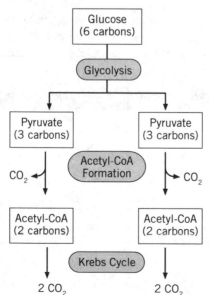

Glucose
(6 carbons)

Glycolysis

Pyruvate
(3 carbons)

Pyruvate
(3 carbons)

CO_2

Acetyl-CoA
Formation

CO_2

Acetyl-CoA
(2 carbons)

Acetyl-CoA
(2 carbons)

Krebs Cycle

2 CO_2

2 CO_2

Carbon flow is the general process of glucose breakdown in cell respiration.

Energy Flow 🔴

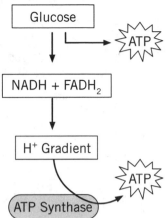

Energy flow is the general process of energy flowing from the bonds in glucose to the bonds in ATP.

ATP 💬

ATP stands for adenosine triphosphate.

Phosphate bonds in ATP store potential energy.

The energy in this bond is used to power the reactions and processes required for cells to survive.

Classifying Organisms in Ecosystems 🔔

Living things are classified by how they obtain food.

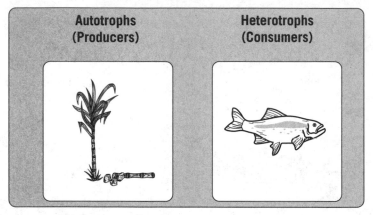

Autotrophs (Producers)	Heterotrophs (Consumers)

- Autotrophs produce complex organic compounds (such as carbohydrates, fats, and proteins) from simple substances in the environment (such as CO_2).
- Most autotrophs use energy from light (photoautotrophs).
- Some autotrophs use inorganic chemical reactions instead (chemoautotrophs).
- Autotrophs serve as primary producers in a food chain. For example, plants are autotrophs.

- Heterotrophs consume other organisms in a food chain.
- A heterotroph is unable to produce organic substances from inorganic ones.
- Heterotrophs depend either directly or indirectly on autotrophs for nutrients and food energy. For example, animals are heterotrophs.

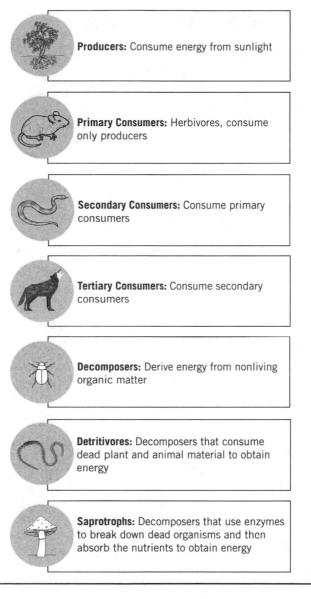

Producers: Consume energy from sunlight

Primary Consumers: Herbivores, consume only producers

Secondary Consumers: Consume primary consumers

Tertiary Consumers: Consume secondary consumers

Decomposers: Derive energy from nonliving organic matter

Detritivores: Decomposers that consume dead plant and animal material to obtain energy

Saprotrophs: Decomposers that use enzymes to break down dead organisms and then absorb the nutrients to obtain energy

 Did You Know?

- Decomposition is important because it returns nutrients to the environment.
- Earthworms, many insects, and crabs are detritivores.
- Bacteria and fungi are saprotrophs.

Food Chains and Food Webs ❗

Energy flows through ecosystems in predictable ways, from the Sun to producers to consumers.

Food chains show how energy flows step by step from producers to consumers. Levels in food chains are called trophic levels. Arrows depict the transfer of energy through trophic levels.

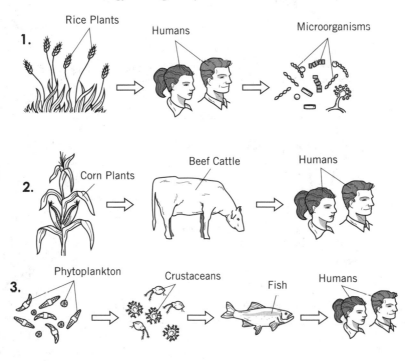

1. Rice Plants → Humans → Microorganisms

2. Corn Plants → Beef Cattle → Humans

3. Phytoplankton → Crustaceans → Fish → Humans

 Remember!

- A single organism may occupy multiple levels of a food chain. When eating sushi, you are a primary consumer because you are eating seaweed and rice, and a secondary consumer by eating fish.
- Herbivores eat plants. Carnivores eat animal tissue. Omnivores eat both.
- Primary producers are always trophic level 1, herbivores are trophic level 2, and predators are trophic levels 3 and higher.

Energy Flow Across Trophic Levels ❗

Energy pyramids show how much energy is available to each successive trophic level. About 10% of energy is passed from one trophic level to the next. The rest (90%) is lost as heat or is used for metabolism and anabolism.

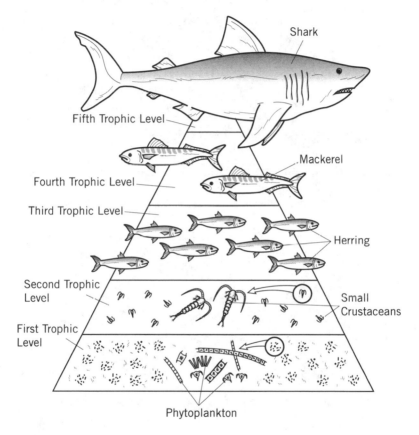

Shark

Fifth Trophic Level

Fourth Trophic Level

Mackerel

Third Trophic Level

Herring

Second Trophic Level

Small Crustaceans

First Trophic Level

Phytoplankton

 Did You Know?

Only 10% of energy transfer per trophic level explains why food chains rarely have more than four trophic levels.

The Living World

Toxins in Food Chains 〰

Energy isn't the only thing that can flow through a food chain. Environmental toxins and heavy metals can also enter food chains.

Bioaccumulation

Many toxins accumulate in an organism because organisms can't break down these molecules.

Biomagnification

Toxin concentration increases with each tropic level of a food chain.

 Ask Yourself...

Why is presence of mercury in fish a particular health concern, and where does this mercury come from?

Food Webs ❗

Whereas food chains show simple relationships, **food webs** show the complex interactions between many species. Food webs also show more realistic feeding relationships.

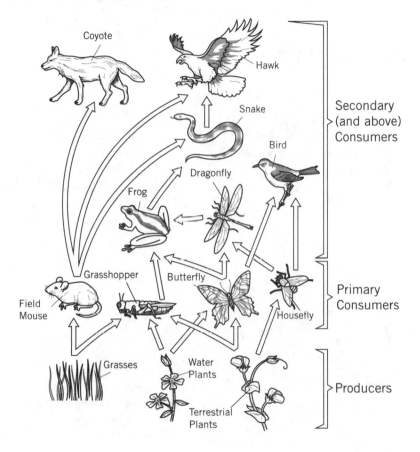

Ecosystem Diversity 🔔

Ecosystem diversity describes how variable an ecosystem is within a geographical location. It impacts human existence and the environment. Ecological diversity is a type of **biodiversity.**

Biodiversity 🔔

The number and variety of organisms found in the world or in a particular habitat or ecosystem

The variability among living organisms (within and between species, within and between ecosystems)

Biodiversity in an ecosystem is a good thing.

More biodiversity means a larger and more diverse gene pool, which leads to a greater chance of adaptation and survival.

Biodiversity in all forms is the result of evolution.

Evolution ❗

Evolution is the change in a population's genetic composition over time. **Natural selection** and **genetic drift** are mechanisms of evolution.

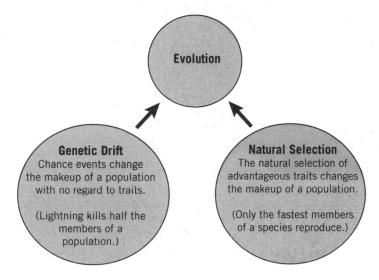

Micro- vs. Macroevolution

Microevolution	When a population displays small-scale changes over a relatively short period of time
Macroevolution	Large-scale patterns of evolution within biological organisms over a long period of time

Natural Selection

- Certain organisms live and reproduce, and others die.
- Beneficial inherited characteristics are passed down to the next generation.
- Unfavorable heritable characteristics become less common in the population.
- This process acts upon a whole population over time, not on one individual organism.

Evolutionary Fitness

An individual with this type of fitness is better adapted to their environment and will live and reproduce, passing their genes onto the next generation.

Charles Darwin is known as the Father of Evolution.

 Did You Know?

The term "survival of the fittest" does not refer to the fastest or the strongest organisms, but rather to those organisms that are able to produce offspring and thus pass on their genetic information.

Phylogenetic Trees 〰

A **phylogenetic tree** is a diagram that shows how organisms are related, based on evolutionary relationships. As shown in the diagram below, lines branch off in different directions and end in a group of organisms with a common ancestor. The branch points represent a hypothetical ancestor whose population split and evolved in two different directions.

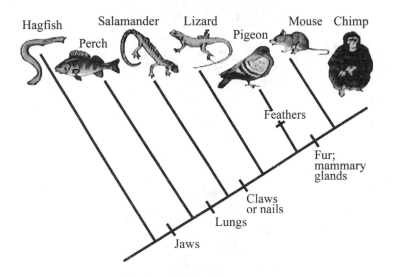

Speciation and Extinction ❗

A **species** is a group of organisms that are capable of breeding with one another but incapable of breeding with other species.

Speciation is the formation of new species from preexisting species. For example, if two populations of a species are separated by a geographic barrier or require different habitats, they can eventually change enough that they become two different species (**genetic drift**).

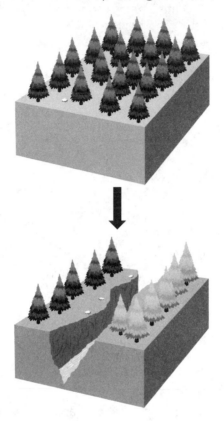

Just as new species are formed by natural selection and genetic drift, other species may become extinct. **Extinction** occurs when a species cannot adapt quickly enough to environmental change and all members of the species die.

Type of Extinction	Definition	Examples
Biological Extinction	• Extermination of a species • No individuals of this species left on the planet	• Dodo bird • Passenger pigeon • Mammoth • Dinosaurs
Ecological Extinction	• So few individuals of a species that this species can no longer perform its ecological function	• Alligators in the Everglades in the 1960s • Wolves in Yellowstone before re-introduction in the 1990s
Commercial or Economic Extinction	• A few individuals exist, but the effort needed to locate and harvest them is not worth the expense	• Groundfish population of the Grand Banks (off the Maritimes of Canada)

Ecosystems can be disrupted or damaged by natural events (such as fire or flooding), or human activities (such as deforestation or overfishing).

Resilience:
How quickly an ecosystem recovers from damage or disturbance

Resistance:
The ability of ecosystems to remain unchanged when subject to disturbance

> **~** If the disturbance is major enough or lasts long enough, a threshold may be reached where the ecosystem undergoes a regime shift, sometimes permanently.

Ecosystem Services ❗

Humans gain many benefits from the natural environment and from properly functioning ecosystems.

Ecosystems are essential to life.

Land, water, and air must be used responsibly if they are to benefit future generations.

Ecosystem Services

The benefits that humans gain from ecosystems can be divided into four categories.

1. Provisioning Services: Physical items we obtain from our environment

Examples: Food, raw materials, water, energy, and medicinal resources (such as plants and mushrooms used to cure health problems)

2. Cultural Services: Non-material benefits people obtain from the ecosystem

Examples: Recreation, science and education, tourism, inspiration for culture and art, and spiritual experience

3. Regulating Services: Benefits obtained from the regulation of ecosystems. These are often hard to see and are taken for granted.

Examples: Pest and disease control, water and air purification, climate regulation, waste decomposition and detoxification, and pollination

4. Support Services: Allow for the other ecosystem services to be present

Examples: Nutrient recycling and soil formation

Natural Ecosystem Change ❗

Just as relationships between individual species are dynamic, characteristics of ecosystems also change over time. Species of plants and animals are continually coming and going, evolving and dying out. Changes can happen over months or even centuries.

❗ In Chapter 7, you'll be learning more about global warming and greenhouse gases. These climate changes can be a driving force in ecosystem change.

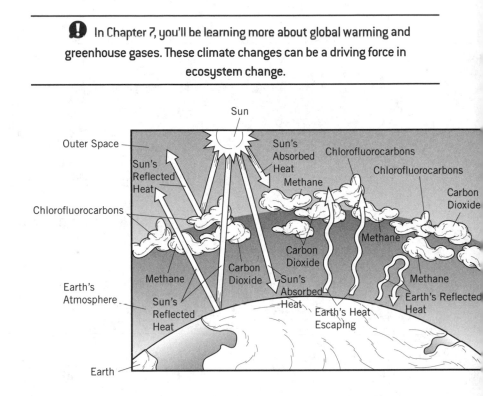

There are several types of species in ecosystems.

Species	Definition	Example	
Keystone	• Maintain the biotic balance in a community • Their presence contributes to an ecosystem's diversity • Their extinction would lead to a large change in the ecosystem	• Fig trees in a tropical forest • Wolves in western North America	❗
Indicator	• Used as a standard to evaluate the health of an ecosystem • More sensitive to biological changes within their ecosystems than other species • Used as an early warning system to detect dangerous changes to a community	Trout in freshwater biomes	💬
Indigenous or Native	• Originate and naturally live in an area or environment	Kangaroos and koalas in Australia	💬
Invasive	• A new species introduced into an environment, by chance, by accident, or with intention • The new species often fails to find a niche and dies, but can sometimes prosper in their new home	• Grey squirrels were introduced to England in 1876, and competed with native red squirrels • In 1904, a fungus was introduced accidentally into the deciduous forests of the eastern U.S., and killed many chestnut trees • Zebra mussels were introduced into the Great Lakes • Kudzu (a vine) was introduced in the U.S. in order to control the problem of erosion	〰

Ecological Succession 🔴

Some of the changes that take place in a geographic area are predictable; this is called **ecological succession,** and it can occur two different ways. The communities in each stage of succession drive the environmental changes that allow the next stage to take over.

Primary Succession

Begins in a virtually lifeless area, such as the area below a retreating glacier

Secondary Succession

Takes place where an existing community has been cleared (such as by fire, tornado, or human impact), but the soil has been left intact

Pioneer Species

• Organisms present in the first stages of either type of succession

• Have wide ranges of environmental tolerance

Climax Community

• Formed in the final stage of succession

• Have a dynamic balance between abiotic and biotic components of the community

 Ask Yourself...

Frogs have shell-less eggs and permeable skins, making them very vulnerable to changes on land and in water quality. What kind of species are they?

Primary Succession ❗

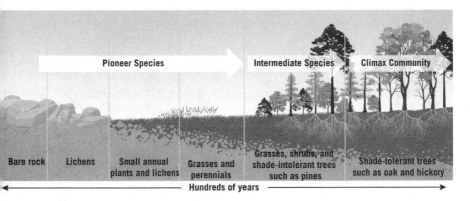

Primary Ecological Succession for a Deciduous Forest

Component	Key Points
Bare rock	Starting material
Lichen, algae, mosses, bacteria	• Pioneer organisms • Break down rock and leave organic debris, which together forms soil
Grasses	Add organic matter to soil and anchor it in place
Small herbaceous plants	Continue to add organic matter to soil
Small bushes and shrubs	Add shelter and shade for other plants and animals
Conifers	Create additional habitats
Short-lived hardwoods such as dogwood and red maple	Can tolerate shade of conifers but are short-lived and vulnerable to damage
Long-lived hardwoods such as such as oak and hickory	More specialized, hardier hardwoods

❗

Secondary Succession

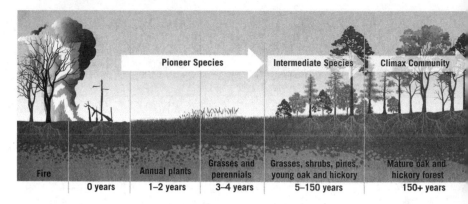

Fire	Annual plants	Grasses and perennials	Grasses, shrubs, pines, young oak and hickory	Mature oak and hickory forest
0 years	1–2 years	3–4 years	5–150 years	150+ years

Remember!

Plant and animal species change through different stages of succession.

Habitat Fragmentation

Habitat fragmentation refers to when a natural habitat is reduced or fragmented, or when a regular and balanced ecosystem is damaged. This causes uneven edge effects and an increase in ecosystem fragility. Fragmentation also negatively affects ecotones and damages the habitat. Refer to the following diagram.

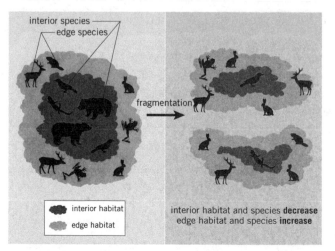

Essential Elements of Life

By mass, living things are:

- 96% O, C, H, and N
- 3.5% Ca, P, K, and Mg
- 0.5% other trace elements (such as Na, Cl, S, Fe, and I)

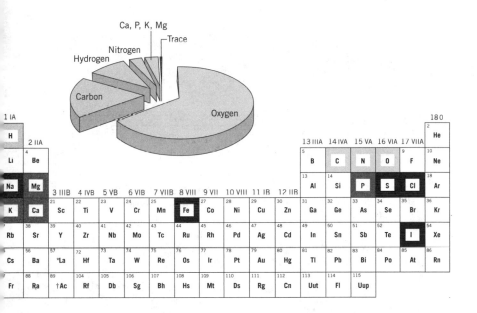

Did You Know?

Earth's atmosphere is made up of about 78% nitrogen and 21% oxygen. The other components are trace elements.

Oxygen
Final electron acceptor in cell respiration and made in photosynthesis

Phosphorus
Used in nucleic acids and some lipids

Carbon
- Backbone of organic molecules

- Found in carbohydrates, proteins, lipids, and nucleic acids

Hydrogen
A small element found in most organic molecules

Nitrogen
Used in proteins and nucleic acids

Natural Biogeochemical Cycles 〰

Nutrients such as carbon, oxygen, nitrogen, phosphorus, sulfur, and water all move through the environment in complex cycles known as **biogeochemical cycles.** These cycles involve living organisms, geologic formations, chemical substances, and various processes and procedures. Energy is provided by the Sun and the Earth; both the core and the mantle provide thermal (heat) energy. Here are some key terms you should keep in your back pocket:

Reservoir

A location where a large quantity of a nutrient sits for a long period of time

Exchange Pool

A site where a nutrient sits for only a short period of time

Residency Time

Amount of time a nutrient spends in a reservoir or an exchange pool

Biotic or Abiotic?

Reservoirs and exchange pools can be abiotic (such as the ocean or a cloud), or biotic (a living organism).

Law of Conservation of Matter

- Matter can neither be created nor destroyed.
- However, nutrients can be made unavailable for cycling (for example, by locking them away in deep ocean sediments).

Water Cycle 〰

Reservoirs	• Lakes • Oceans • Long-term snow and ice
Exchange Pools	• Living organisms • Groundwater • Runoff water • The atmosphere
Processes	• Precipitation (rain): gas to liquid, falls due to gravity • Evaporation: liquid to gas • Transpiration: water evaporates out of plant leaves

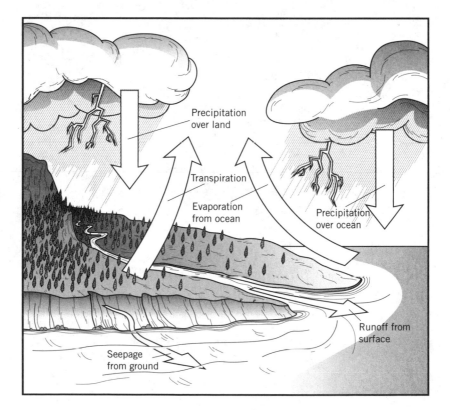

Precipitation over land

Transpiration

Evaporation from ocean

Precipitation over ocean

Runoff from surface

Seepage from ground

Carbon Cycle 🌫

Reservoirs	• Oceans • Earth's rocks • Fossil fuels
Exchange Pools	• Living organisms • The atmosphere
Processes	• Cell respiration: carbohydrates to CO_2 • Photosynthesis: CO_2 to carbohydrates • Decomposition: organic matter to CO_2 • Consumption via the food chain: C transfers from organism to organism

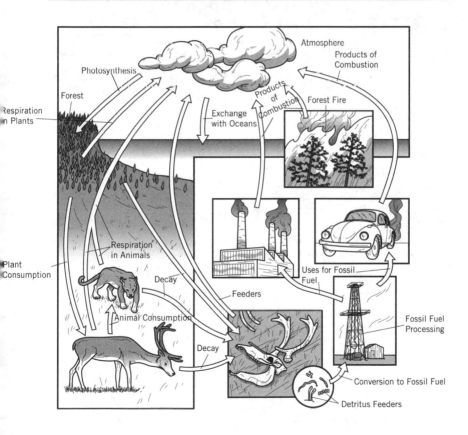

CO_2 is soluble in water. This is how carbonated beverages such as sodas are made.

The Living World

Nitrogen Cycle ∽

Reservoirs	The atmosphere
Exchange Pools	• Living organisms • Soil • Groundwater
Processes	• Fixation: N_2 to ammonia (NH_3) or nitrate (NO_3^-) • Nitrification: NH_3 or ammonium (NH_4^+) to nitrite (NO_2) • Assimilation: plants absorb NH_3, NH_4^+, and NO_3^- through their roots • Consumption via the food chain: N transfers from organism to organism • Ammonification: decomposers break organic matter down into NH_3 or NH_4^+ • Denitrification: NH_3 to NO_3^- or NO_2, then to N_2 or nitrous oxide (N_2O)

 Did You Know?

• Nitrogen fixation makes atmospheric nitrogen available to living organisms. It is mostly performed by soil bacteria such as *Rhizobium*.
• Anaerobic bacteria perform most denitrification.

The Nitrogen Cycle

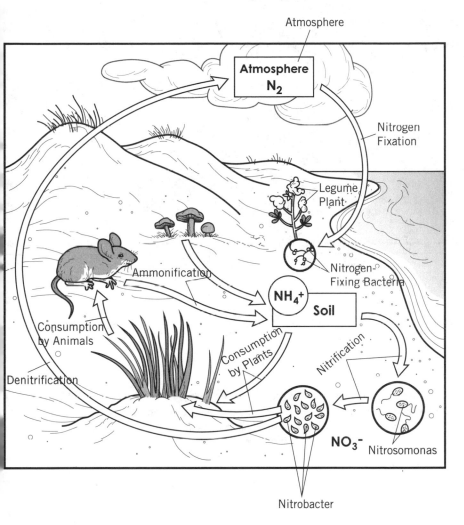

Phosphorus Cycle

Reservoirs	• Rock • Soil • Sediments
Exchange Pools	• Living organisms • Groundwater • Runoff water
Processes	• Chemical weathering: P in reservoirs is released as phosphate (PO_4^{3-}) • Plants absorb soil PO_4^{3-}; *Mycorrhizae* fungi can help • Mining: P-rich rocks are used to produce fertilizer • Leaching: fertilizers to ground or runoff water, then aquatic ecosystems

 Did You Know?

• Phosphorous only exists in the atmosphere in dust particles.
• Eutrophication occurs when a body of water receives excess nutrients. This can cause overgrowth of algae and deplete the water of oxygen.

Remember!

A limiting factor is any factor that controls a population's growth. Phosphorous is often a limiting factor for plant growth; plants that have little phosphorous are stunted.

The Phosphorus Cycle

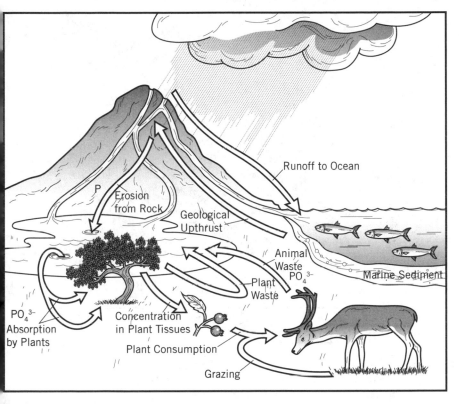

Runoff to Ocean

P Erosion from Rock

Geological Upthrust

Animal Waste

PO_4^{3-}

Plant Waste

Marine Sediment

PO_4^{3-} Absorption by Plants

Concentration in Plant Tissues

Plant Consumption

Grazing

Sulfur Cycle ～

Reservoirs	• Rocks • Salts • Ocean sediments
Exchange Pools	• Living organisms • The atmosphere
Processes	• Plants absorb S from the soil via roots • Consumption via the food chain: S transfers from organism to organism • Volcanic eruptions release SO_2 • Decomposition and decay: organic matter to S • Human activities: sulfur dioxide (SO_2) and hydrogen sulfide (H_2S) released into atmosphere

 Did You Know?

Sulfur is required to make proteins and vitamins, so plants and animals both need sulfur in their diets.

The Sulfur Cycle

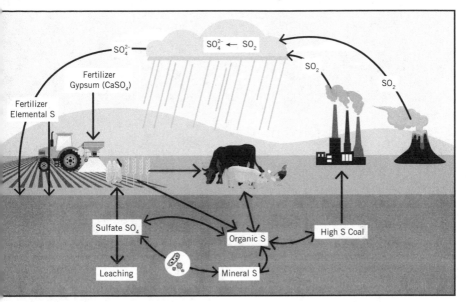

Chapter 2 Summary

- Organization of living things can be thought of as:

Biosphere → Ecosystems → Communities → Populations (of species)

- Three main types of relationships govern interactions between species: competition, predation, and symbiosis.
- Ecosystems are divided into biomes (land) and zones (water). There are many different types of terrestrial biomes and aquatic zones.
- Energy is passed through ecosystems via food chains and food webs.
- Biodiversity is the result of speciation and natural selection of traits. Extinction allows for niches to become available and new traits or species to evolve to use those niches.
- The five major cycles you should know are water, carbon, nitrogen, phosphorus, and sulfur.

CHAPTER 3
Population

In this chapter, we'll start by discussing some important characteristics of populations, and then lead you through a short section on how and why populations grow. Next we'll get to the heart of the topic in a section specifically devoted to human population growth.

Population Ecology ❗

Key Concepts 💬

A **population** is defined as a group of organisms of the same species that inhabits a defined geographic area at the same time. Individuals in a population generally breed with one another, rely on the same resources to live, and are influenced by the same factors in the environment.

Two important characteristics of populations are

- **Population density:** The number of individuals of a population that inhabit a certain unit of land or water area. An example of population density would be the number of squirrels that inhabit a particular forest.
- **Population dispersion:** How individuals of a population are spaced within a region. There are three main ways in which populations can be dispersed: *clumped, uniform,* and *random.*

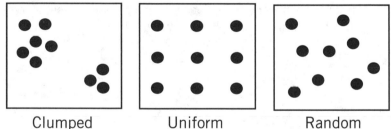

    Clumped   Uniform   Random

😎 What city has the largest rodent population? Hamsterdam!

	Clumped	Uniform	Random
Definition	The most common dispersion pattern for populations, which adheres to the "those of a feather flock together" idea	Members of a population are uniformly spaced throughout their geographic region; usually a result of competition for an ecosystem's resources.	A relatively uncommon pattern of dispersion, random dispersion occurs when the position of each member of a population is not determined or influenced by other members of the population.
Example(s)	Plants grow together in a region that suits their requirements for life; fish swim in schools to avoid predation; birds migrate in groups.	Trees are uniformly distributed in forests so that each receives adequate light and water.	The locations of plants interspersed in a field are random and relative to other species outside of their population.

 Ask Yourself...

Emperor penguins are highly social animals that live and feed in large groups. Would populations of emperor penguins likely be clumped, uniformly spaced, or randomly dispersed?

Population Growth and Carrying Capacity 😧

Biotic Potential 💬

- How much a population would grow if there were unlimited resources in the environment
- Not a practical model for population growth because, in reality, resources in an environment are limited

Carrying Capacity 😧

- The maximum population size that can be supported by the available resources in the region
- Differs by species, as different species have different requirements for life
- Example: A population of bacteria is larger than a population of zebras in a given area; bacteria are much smaller than zebras and, therefore, require fewer resources for survival.

Graphing Population Growth 😧

If we were to plot the growth of a population of bacteria over time, the curve produced would be in the shape of a J because the bacteria grow exponentially. The **J-Curve** models such exponential growth, as shown by the following graph.

Exponential (Unrestricted) Growth

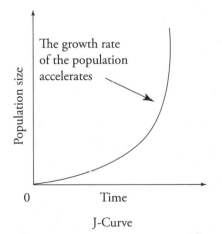

J-Curve

However, such exponential growth only occurs when a population has access to unlimited resources. Because such ideal settings are rare and fleeting in nature, the J-Curve is not the ideal model of population growth in the natural environment.

In a more realistic model for population growth, after the initial burst in population, the growth rate generally drops, and the curve ultimately resembles a flattened *S*. This type of growth, which is a much better model for what exists in natural settings, is called **logistic population growth**.

Logistic (Restricted) Growth

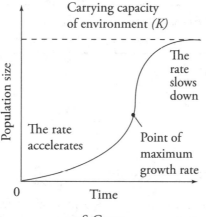

S-Curve

The logistic growth model basically says that when populations are well below the size dictated by the carrying capacity of the region they live in, they will grow exponentially; but as they approach the carrying capacity, their growth rate will decrease and the size of the population will eventually becomes stable.

Rule of 70 💬

We can predict long-term population growth using a model called the Rule of 70. The **Rule of 70** states that the time it takes for a population to double can be approximated by dividing 70 by the current growth rate of the population. For example, if the growth rate of a population is 5 percent, then the population will double in 14 years: $\frac{70}{5 \text{ percent}} = 14$ years.

The Rule of 70 can be used to estimate the number of years for any variable to double, the doubling time.

Ask Yourself...

If the current growth rate of a population is 25%, how many years will it take for the population to double?

Reproductive Strategies ❗

The rate of growth of a population depends on the species that makes up the population. Species can be divided into two groups based on their reproductive strategies: the *r*-**selected** pattern or the *K*-**selected** pattern.

r-Selected Organisms	*K*-Selected Organisms
• Reproduce early in life • Usually have a high capacity for reproductive growth • Little or no care given to offspring, but due to sheer numbers of offspring in the population, enough will survive to allow population to continue • Examples: bacteria, algae, protozoa	• Reproduce later in life • Produce fewer offspring • Devote significant time and energy to nurturing of offspring—it is important to preserve as many members of the offspring as possible because they produce so few • Parents are invested in each individual offspring • Examples: humans, lions, cows

Many species lie on the continuum between these two strategies, but the groups are useful for broad comparisons.

Survivorship and Population Cycles ❗

When we observe populations in their natural habitats, there are two distinct patterns that occur: the **boom-and-bust cycle** and the **predator-prey cycle.**

Boom-and-Bust Cycle ❗

- Very common among r-strategists
- Characterized by rapid increase in the population and then an equally rapid drop-off; this may be linked to predictable cycles in the environment (e.g., temperature or nutrient availability)
- When conditions are good for growth, the population increases; when conditions worsen, the population's numbers decline.

The following graph shows this type of cycle in action.

Boom-Bust Cycle

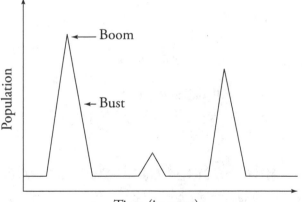

Predator-Prey Cycle !

Here is the basic idea behind the predator-prey cycle, using the example of rabbits and coyotes: In a year of relatively high rainfall, rabbits have plenty of food, which enables them to reproduce very successfully. In turn, because the coyote is a predator of the rabbit, coyotes would also have plenty of food, and their populations would also rise rapidly. However, if the rainfall is below average a few years later, then there would be less grass, the population of rabbits would decline, and the coyote population would decline in turn. The graph of the predator-prey relationship looks like the following.

Predator-Prey Cycle

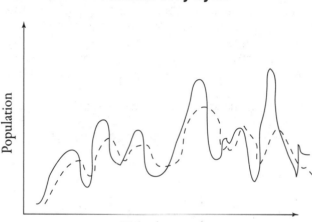

Time (in years)

———— = Population of rabbits (prey)

- - - - = Population of coyotes (predator)

Factors Influencing Population Growth !

Density-dependent factors are population-limiting factors that are purely the result of the size of the population. These factors include

- increased predation
- competition for food or living space
- disease (which can spread more quickly in overcrowded populations)
- buildup of toxic materials

Density-independent factors are population-limiting factors that operate independently of population size. Some of these are fires, storms, earthquakes, and other catastrophic events.

The **survivorship curve** indicates the probability that a given organism will live to a certain age, taking into account the various factors that influence population growth. The survivorship curve accounts for three types of survivorship:

- **Type I (K-selected):** The majority of offspring will live for a long period of time; eventually they will start to die off.
- **Type II:** Offspring have a 50-50 chance of surviving to old age.
- **Type III (r-selected):** Most offspring die young, but if they live to a certain age, they will live a longer life.

Survivorship Curve

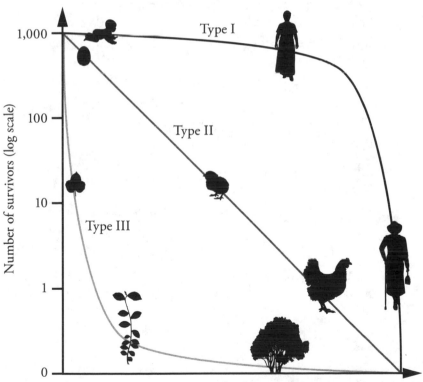

Age in relative units

Human Population ❗

World Population ❗

According to the World Population Clock, the world population as of March 2018 is estimated to be 7.6 billion. The birth rate has actually fallen in the United States and worldwide, but the world population is still increasing. Take a look at the following table, which shows esti mated populations of some major countries as of January 2018.

Top 10 Most Populous Countries	
Country	Estimated Population (January 2018)
China	1,384,688,986
India	1,296,834,042
United States	329,256,465
Indonesia	262,787,403
Brazil	208,846,892
Pakistan	207,862,518
Nigeria	195,300,343
Bangladesh	159,453,001
Russia	142,122,776
Japan	126,168,156

Source: U.S. Census Bureau, International Data Base

Ask Yourself...

The current growth rate is approximately 1.10% per year, or 83 million people annually. At this rate, when will the world population reach 10 billion?

The following graph shows how the overall world population of humans has increased since 1950, and predicts it will continue to increase into the 2050s.

World Population: 1950–2050

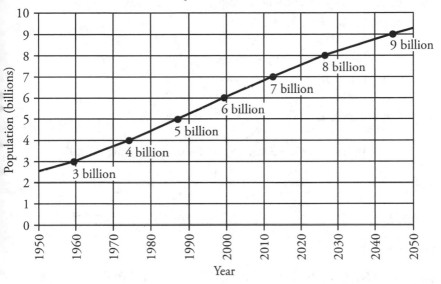

Source: U.S. Census Bureau, International Data Base, December 2010 Update

We can determine the rate of population change of a country by using a simple formula, if we only consider the contributions of births and deaths to changes in population size.

❗ **Birth rate*** = the number of live births per 1,000 members of the population in a year

Death rate = the number of deaths per 1,000 members of the population in a year

Actual Growth Rate (%) =

$$\frac{(\text{birth rate} + \text{immigration}) - (\text{death rate} + \text{emigration})}{1,000}$$

The table below shows the 2017 **growth rates** for 11 countries and the 2015 birth and death rates. Note that some countries are experiencing negative growth.

Country	Growth Rate % Per Year	Birth Rate Per 1,000	Death Rate Per 1,000
United States	0.71	12	8
Japan	–0.21	8	10
China	0.43	1	8
Russia	0.02	12	14
Latvia	–1.06	12	7
Indonesia	1.10	17	6
Canada	.92	10	8
Mexico	1.27	19	5
Nigeria	2.63	21	10
India	1.13	20	7
United Kingdom	0.6	12	9

Source: Worldometers and the U.S. Census Bureau, International Data Base

---

*Birth rate is sometimes referred to as crude birth rate; likewise, death rate is sometimes called crude death rate.

Population Changes

Populations can also change in number as a result of migration into and out of the population. Two important terms to describe human migration are

- **emigration**, the movement of people *out of* a population
- **immigration**, the movement of people *into* a population

Keep in mind that, in general, emigration and immigration are only small factors in the changes in size of human populations.

The most significant additions to human populations are due to births, plain and simple.

Total Fertility Rate (TFR)	• The number of children a woman in a given population will bear during her lifetime • Based on an analysis of data from preceding years • While they provide a rough estimate, TFRs cannot be depended on because they assume the conditions of the past will be the conditions of the future.
Replacement Birth Rate	• The number of children a couple must have in order to replace themselves in a population • Worldwide, the replacement birth rate is slightly higher than 2. • In developing countries, this number is as high as 3.4 due to higher mortality rates.

❗ Factors Affecting the Total Fertility Rate of a Population

- Availability of birth control
- Demand for children in the labor force
- Base level of education for women
- Existence of public and/or private retirement systems
- The population's religious beliefs, culture, and traditions

---

- Base level of education for women
- Dominant religion
- Base level of education for women
- Existence of public and/or private retirement systems
- The population's religious beliefs, culture, and traditions

 Did You Know?

The reason that religion and culture are predictors of birth rates is
that in some countries, certain groups have a proclivity
toward reproduction for religious reasons.

The world's population has grown considerably, especially in the past 100 years, because of the significant drop in the world death rate (as opposed to an increased birth rate). People are living longer lives, and there are far fewer infant deaths today than there were 100 years ago. This is due, in large part, to the Industrial Revolution, which improved the standard of living for millions living in industrialized nations. Other causes of the extension of the human life span are

- development of clean water sources
- better sanitation
- the creation of dependable food supplies
- better health care

In general, the overall health of a population can be estimated by
examining the expected life span of individuals and the mortality
rate of infants.

Age-Structure Pyramids ❗

Age-structure pyramids (also called **age-structure diagrams**) are useful for graphically representing populations. Some age-structure diagrams group humans into three age categories:

- **Pre-reproductive** (0–14 years)
- **Reproductive** (15–44 years)
- **Post-reproductive** (45 years and older)

Age-structure pyramids group members of the population strictly by age, with each decade representing a different group. The *x*-axis contains the information relating to the percent or number of individuals in each of the age groups.

Age-Structure Pyramid

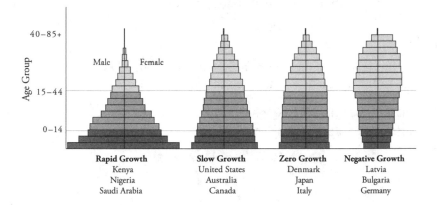

Age-structure pyramids can be used to predict population trends. For example, when the majority of a population is in the post-reproductive category, the population size will decrease in the future because most of its members are incapable of reproducing.

The opposite is true if the majority of a population is in the pre-reproductive category; these populations will increase in size as time goes on. For example, as shown in the previous graph, Nigeria has a large number of pre-reproductive and reproductive members in its population, while the United States has a fairly even distribution. From this, we can see that the population of Nigeria should increase significantly over time—it has what is referred to as **population momentum**—while the population of the United States should grow more slowly.

You should be able to identify the growth rate of a country based upon its age-structure pyramid.

Demographic Transition Model ❗

The **demographic transition model** is used to predict population trends based on the birth and death rates of a population. In this model, a population can experience zero population growth via two different means:

1. as a result of high birth rates and high death rates
2. as a result of low birth rates and low death rates

When a population moves from the first state to the second state, the process is called **demographic transition.**

The four states that exist during this transition are the following:

1. **Preindustrial state**
   - Population exhibits slow growth rate, and has a high birth rate and high death rate due to harsh living conditions.
   - Harsh living conditions can be considered environmental resistance, an umbrella term for conditions that slow a population's growth.
2. **Transitional state**
   - Birth rate is high, but death rate is lower due to better food, water, and health care.
   - This allows for rapid population growth.
   - Birth rates remain high due to cultural or religious traditions and a lack of education for women.
3. **Industrial state**
   - Population growth is still fairly high, but the birth rate drops, becoming similar to the death rate.
   - Many developing countries are currently in the industrial state.
4. **Postindustrial state**
   - Population approaches and reaches a zero growth rate.
   - Populations may also drop below the zero growth rate.

Demographic Transition Model

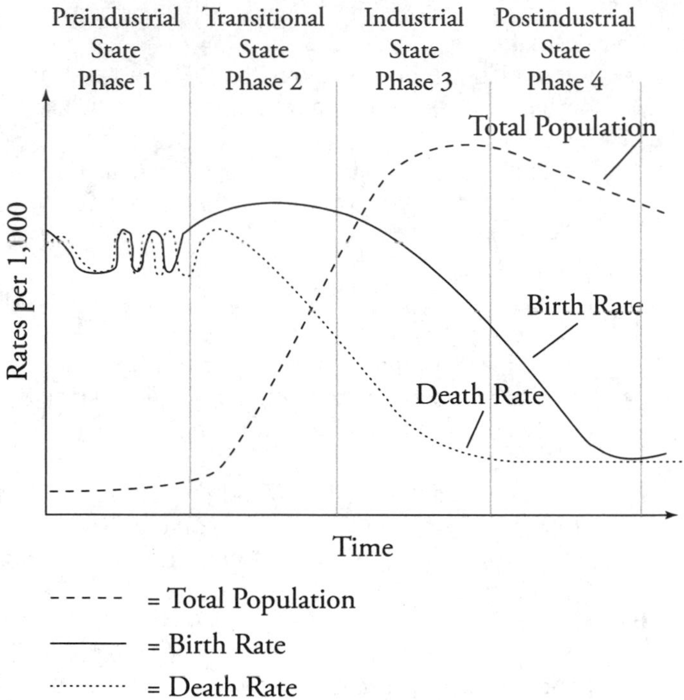

| | Preindustrial State Phase 1 | Transitional State Phase 2 | Industrial State Phase 3 | Postindustrial State Phase 4 |

Total Population

Birth Rate

Death Rate

Rates per 1,000

Time

- - - - - = Total Population
———— = Birth Rate
·········· = Death Rate

Human Impact on Earth ❗

Humans have the greatest impact on the environment of any living species on Earth. The increase in the human population over the last few centuries has seriously and dramatically changed the face of the Earth.

Ecological Footprint 💬

An **ecological footprint** is used to describe the environmental impact of a population. It is defined as the amount of Earth's surface that's necessary to supply the needs of, and dispose of the waste of, a particular population. Americans have one of the largest ecological footprints; we require about 9.7 hectares per capita (per person).

We can use a mathematical model to describe the impact that humans have on the environment. Nicknamed the **IPAT model**, it is written as

$$I = P \times A \times T$$

In the model, I = the total impact, P = population size, A = affluence, and T = level of technology.

Urban Areas 💬

Almost half of the world's population today lives in an urban area. In the United States, this is partly due to the fact that the aging population has largely moved into the cities to have greater access to health services, employment opportunities, and cultural activities. When considering those who live in urban areas, we also count those who reside in satellite communities, or **suburbs**.

The term used to describe the emigration of people out of the city and into the suburbs is **urban sprawl**. In some areas of the United States, urban sprawl takes over vast tracts of land.

In general, affluent populations have a much higher ecological footprint than non-affluent ones.

Urban Sprawl

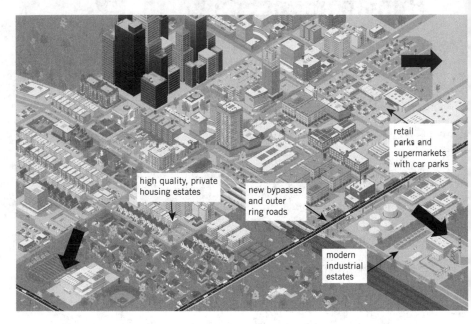

high quality, private housing estates

new bypasses and outer ring roads

retail parks and supermarkets with car parks

modern industrial estates

Problems Associated with Urban Sprawl 💀

While many people find the suburbs a pleasant place to live, ecologists and city planners have recently come to realize that urban sprawl may reduce quality of life for all urban dwellers in the form of, among other things, increased water usage and waste.

Solutions: Building Sustainable Cities 💬

A current trend in urban planning and design is **sustainable design,** which is the philosophy of designing physical objects, buildings, and homes to comply with the principles of economics and ecological sustainability.

Environmental Problems ⚠️

With the **Green Revolution** that occurred in the 1950s and 1960s came technological innovation and an increased use of pesticides and fertilizers, which allowed farmers to increase crop production throughout the world. However, the enormous amount of food production takes its toll on the land, as exemplified in problems like overgrazing, desertification, irrigation agriculture, and deforestation.

Overgrazing and Desertification

Extensive pastoralism, the shifting of animal herds between grazing pastures, has remained popular in several arid parts of the world (especially Africa, Middle East, Central Asia), where dry grassland is the common landcover. This problem is similar to that of rainforest destruction, since too many people and too many animals are placing **population pressure** on too little land. **Overgrazing** has led to significant amounts of dry grassland being denuded, eroded, and as a result, desertified. **Desertification** is any human process that turns a vegetated environment into a desert-like landscape. In addition to overgrazing, deforestation and **soil salinization** can also lead to desertification.

Irrigation Agriculture

The practice of **irrigation** opens up more land to cultivation than would normally be possible in arid climates. Irrigation agriculture is responsible for close to three-quarters of world freshwater use and up to 90% of freshwater use in the most poverty-stricken countries of the world. Governments often heavily subsidize irrigation agriculture and the crops produced are often worth less than the water. The Nile Valley in Egypt is an example of heavily subsidized irrigation agriculture. Unfortunately, the water for these irrigation farms comes from underground water tables called **aquifers.** These aquifers are being depleted at a rapid rate and large-scale grain-producing countries such as India, China, and the United States are examples of those caught in this predicament.

Deforestation

Deforestation is the act of clearing, generally by cutting down or burning, a forested area of trees and other vegetation without the intention of replanting. Not only does rainfall sap the remaining nutrients from the soil, but deforestation places pressure on a valuable, natural resource—oxygen—and lessens the Earth's ability to filter CO_2.

Another problem that arises from the large number of grazing animals worldwide is the large amount of **animal waste** produced. Manure is not used as fertilizer due to difficulty with transport. It has instead become the most widespread source of water pollution in America.

Limited Resource Availability and Hunger ❗

〜 Key Terms and Concepts

Macronutrients—Nutrients needed in large amounts; e.g., proteins, carbohydrates, and fats

Micronutrients—Nutrients needed in smaller amounts; e.g., vitamins, iron, and minerals such as calcium

Hunger—Occurs when insufficient calories are taken in to replace those that are being expended

Malnutrition—Poor nutrition that results from an insufficient or poorly balanced diet; people whose diets lack essential vitamins and other components often suffer from it

Undernourished—Not meeting a quality of nourishment to sustain proper health and growth. According to the Food and Agriculture Organization, or FAO, 795 million people on Earth are undernourished. Some 780 million of these people are living in developing countries, but, perhaps surprisingly, the remaining 15 million are living in developed nations.

 Did You Know?

〜 Despite its high obesity rate, there are hungry people even in the United States, one of the richest countries in the world. According to the United States Department of Agriculture, 15.6 million Americans were considered "food insecure" and at least one in seven people in the United States relied on food stamps in 2016. The main reason is poverty. Many neighborhoods in which the majority of the citizens that reside there are low-income are often called food deserts. This is because access to fresh, healthy food is difficult. The residents rely on low-quality processed foods for subsistence.

U.S. Households with Children by Food

Food-insecure households—16.5%

Food insecurity among adults only in households with children—8.5%

Food-insecure children—8.0%

Low food security among children—7.2%

Very low food security among children—0.8%

Food-secure households 83.5%

Source: USDA, Economic Research Service, using data from the 2016 Current Population Survery Food Security Supplement

Threatened and Endangered Species ❗

Another way humans impact Earth is through their interaction with animals. Human activities have caused or contributed to the **extinction** of many species. The International Union for Conservation of Nature (IUCN) evaluates the status of plant and animal species. Threatened species are assigned to one of the following three categories:

1. Vulnerable	The species is likely to become endangered if no action is taken.
2. Endangered	The species is likely to become extinct.
3. Critically Endangered	The species is at a very high risk of extinction.

The IUCN maintains a Red List of Threatened Species that is updated regularly.

IUCN Red List:

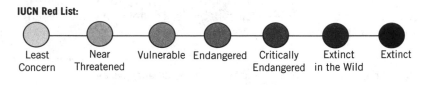

| Least Concern | Near Threatened | Vulnerable | Endangered | Critically Endangered | Extinct in the Wild | Extinct |

In 2000, the number of threatened species was just over 10,000. As of 2017, this number has surpassed 25,000.

Total Threatened Species, 2002–2017

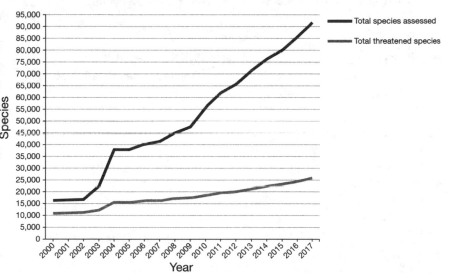

Extinction has occurred throughout Earth's history. This natural rate of extinction is called the **background extinction rate**. Knowledgeable scientists estimate that the current extinction rate is between 50 and 500 times higher than in the past, probably due to human influence. The species that are most endangered have several factors in common. They

- require large ranges of habitat to survive.
- have low reproductive rates.
- have specialized feeding habits.
- have some sort of value to humans (medicinal or food).
- have low population numbers.

Humans play a major role in the extinction of species because of our destruction of animal and plant habitats. Poverty and rapid population growth cause people to use destructive practices, such as slash-and-burn farming, that destroy species' habitats. The following are key terms and concepts related to this process.

! **Fragmentation**—Habitats are broken into smaller pieces by, for example, building of roads and cities.

Degradation—Pollutants are added to the environment.

Overexploitation of animal products—Along with direct hunting, this can contribute to extinction.

Biodiversity hotspot—A term coined by Dr. Norman Myers to describe a highly diverse region that faces severe threats and has already lost 70 percent of its original vegetation.

Use the acronym **HIPPCO** to memorize the causes of extinction.
- **H**abitat destruction/fragmentation
- **I**nvasives
- **P**opulation
- **P**ollution
- **C**limate change
- **O**verharvesting/overexploitation

There are many United States laws that have been passed to reduce the rates of extinctions and protect specific organisms. Three very important ones are the following:

Date	Name of Legislation	What It Did
1972	Marine Mammal Protection Act	This act protected marine mammals from falling below their optimum sustainable population levels.
1973	Endangered Species Act	The act prohibited the commerce of those species considered to be endangered or threatened.
1973	Convention on International Trade in Endangered Species of Wild Flora and Fauna (CITES)	This agreement bans the capture, exportation, or sale of endangered and threatened species.

Chapter 3 Summary

- Population patterns: The distribution of a population is influenced by the density a species can handle based on its resource use and interspecific competition.
- Limits in the ecosystem define the maximum population size, known as the carrying capacity (K).
- The rate of growth of the population depends on the species: K-selected species or r-selected species.
- Human populations: Though birth rates are falling, the population is still increasing.
- Be able to read an age-structure pyramid!
- The demographic transition model is used to predict population trends based on birth and death rates of a population.
- Increasing human population on Earth has led to environmental effects such as resource depletion, hunger and malnutrition, threatened and endangered species, species extinction, and development of suburbs and urban sprawl.
- Key legislation to deal with such environmental problems include the Marine Mammal Protection Act (1972) and the Endangered Species Act (1973).

CHAPTER 4

Land and Water Use

This chapter describes the resources of the world, renewable and nonrenewable, including those from agriculture, forests, oceans, and mines. It will also go a bit into the economics behind our resource use.

Resource Utilization 〔!〕

A **resource** is strictly defined as an available supply that can be drawn on as needed. Published by Garrett Hardin in 1968, **"The Tragedy of the Commons"** often comes to mind when discussing the management of common property resources such as air, water, and land.

The Tragedy of the Commons

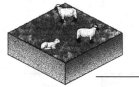 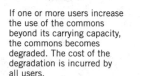

Use of the commons is below the carrying capacity of the land. All users benefit.

If one or more users increase the use of the commons beyond its carrying capacity, the commons becomes degraded. The cost of the degradation is incurred by all users.

Unless environmental costs are accounted for and addressed in land use practices, eventually the land will be unable to support the activity.

"Each [person] is locked into a system that compels him to increase his herd without limit—in a world that is limited. Ruin is the destination toward which all [people] rush, each pursuing his own best interest in a society that believes in the freedom of the commons. Freedom in a commons brings ruin to all."

—Garrett Hardin

The "Tragedy of the Commons" serves as a foundation for modern conservation. **Conservation** is the management or regulation of a resource so that its use does not exceed the capacity of the resource to regenerate itself.

This is different from **preservation**, which is the maintenance of a species or ecosystem in order to ensure its perpetuation, with no concern as to its potential monetary value.

In this chapter, we'll continue to show how human economics often influences how we interact with Earth's resources.

Renewable and
Nonrenewable Resources ❗

When we describe something as a resource, we are essentially putting an economic value on it; therefore, natural resources are described in terms of their value as **ecosystem capital** or **natural capital.**

There are two main types of resources:

---

1. **Renewable resources** are those that can be regenerated quickly. The time necessary for hardwood trees to mature (about 50 years) is widely considered the crossover point from renewable resources to nonrenewable resources.

2. **Nonrenewable resources** are things like minerals and fossil fuels. Nonrenewable resources are typically formed by very slow geologic processes, so we consider them incapable of being regenerated within the realm of human existence.

The **production** refers to the harvest and use of environmental resources for profit, while the **consumption** of natural resources refers to the day-to-day use of environmental resources such as food, clothing, and housing.

Perhaps our greatest resource consumption is towards the production of electricity. So, where do we get the energy that we use to heat up that water in the first step of the creation of electricity?

 Ask Yourself...

Consider coal, oil, wind, and natural gas. Which of these energy sources do you think contributes least to global warming?

The three main sources for electricity production in the United States are **fossil fuels, nuclear energy,** and **renewable energy.** We will discuss our energy resources in more depth later in this book, but here's a quick preview:

Fossil Fuels	Nuclear Energy	Renewable Energy
• Provide 65% of the world's electricity • Formed from the fossilized remains of once-living organisms; over time, this organic matter was exposed to intense heat and pressure, eventually breaking the organic molecules down into oil, coal, and natural gas	• The world's primary non-fossil fuel, nonrenewable energy source • In the U.S., 20% of electrical energy is provided by nuclear power plants. • Worldwide, more than 400 nuclear power plants produce approximately 13% of the world's electrical energy. • China and India lead the world in the creation of new nuclear facilities.	• Bottomless energy source = advantage • Examples include biomass, solar, wind, geothermal, and hydropower energy. • Globally, only about 15% of our energy needs are currently met using renewable energy sources. • Energy storage limitations hinder the expansion of renewable energy

Why is wind power so popular? Because it has a lot of fans!

ASAP Environmental Science

Agriculture ❗

Agricultural revolutions throughout history have resulted in important farming innovations that reduced the labor needed to produce goods, as well as increased the goods harvested per unit of land.

Key Agriculture Terms and Concepts

Traditional subsistence agriculture: Each family grows crops for themselves, relying primarily on animal and human labor for the planting and harvesting. This process provides enough food for the family's survival.

Intensive mixed farming: Farming that allows people to settle permanently and subsist without having to migrate seasonally; traditional subsistence agriculture is an example of this.

Extensive subsistence agriculture: Farming that occurs when there are low amounts of labor inputs per unit of lands.

 Ask Yourself...

What do you think caused these different forms of agriculture to arise? How would geography play a role in the type of agricultural practices used?

Agricultural Innovations 💬

The **First Agricultural Revolution**, also known as the **Neolithic Revolution**, refers to the period of time during which people transitioned from hunting and gathering for food to an organized system of farming.

Vegetative planting, where shoots, stems, and roots of existing wild plants are collected and grown together, was the first phase of early farming.

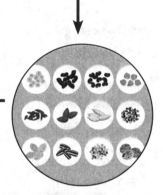

This led to **seed agriculture,** where fertilized seed grains and fruits of plants are collected and replanted together.

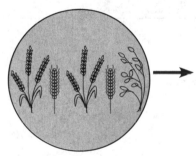

Over time, farmers rejected poorly growing crops and took cuttings and seeds from productive, heartier crops to grow future generations. This **domestication of plants** led to early forms of **horticulture.**

Animal domestication, the process in which selected animal populations become accustomed to human provision and control, was established as an alternative to hunting and fishing.

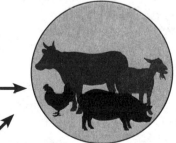

Productive breeds were purposely interbred or hybridized for reproduction through a process known as **animal husbandry.** The diffusion of animal hybrids was specialized by region.

Did You Know?

The **Second Agricultural Revolution** was characterized by technological innovations in agriculture and manufacturing that drastically reduced labor requirements and increased the scale of farm production. This revolution introduced

- specialized **hybrid crops**
- **artificial chemical fertilizers**
- early **chemical pesticides**
- mechanization
- tractors, combine harvesters, and other mechanized food processing devices, which made vast improvements in crop yields for farmers

The **Green Revolution**, which was mentioned in Chapters 1 and 3, occurred in the 1950s and 1960s and is generally thought of as the time after the Industrial Revolution when farming became mechanized and crop yields in industrialized nations boomed. Such innovations also allowed famers in the Third World to increase crop production on small plots of land. The chart below shows a short comparison of the positive and negative effects of the Green Revolution on agriculture and the landscape.

Positive Effects	Negative Effects
• Mechanization, fertilizers, and pesticides that led to improved crop production on small plots of land and a reduction in world hunger • Decreased food costs • Affordable **irrigation pumps** that move water to farming regions • **Genetic engineering** allowing farmers to modify crops with beneficial traits and create **high-yield seeds** • **Biotechnology** research resulting in the development of vaccines, antibiotics, and growth hormones	• Suffering small farms due to their inability to capitalize on **economies of scale** • Detrimental environmental impact caused by the increased use of energy, materials, and machinery • Chemical use leading to **eutrophication,** when fertilizer flows into bodies of water, floods the water with nutrients, and causes explosive algae growth and water pollution • Use of fertilizers and pesticides resulting in **pesticide resistance** • **Genetic engineering** resulting in distrust

Agricultural Activity and the Environment ❗

Irrigation ❗

Problem	How Problem Is Being Addressed	Drawbacks
Repeated irrigation can cause serious problems, including a significant buildup of salts on the soil's surface, which makes the land unusable for crops.	To combat this **salinization** of the land, farmers have begun flooding fields with massive amounts of water in order to move the salt deeper into the soil.	• The large amounts of water can waterlog plant roots, which will kill the crops; this process also causes the water table of the region to rise. • The water for these irrigation farms comes from underground water tables called **aquifers,** which are being depleted at a rapid rate.

Monoculture ❗

In a **monoculture**, just one type of plant is planted in a large area. **Plantation farming** is a type of industrialized agriculture in which a monoculture cash crop is grown and then exported to developed nations. However, a lack of genetic variation makes crops more susceptible to pests and diseases. Additionally, the consistent planting of one crop in an area eventually leaches the soil of nutrients.

 Ask Yourself...

Why would continually planting a single crop have a detrimental impact on the land? How could these programs be overcome?

Prevention of Soil Degradation ❗

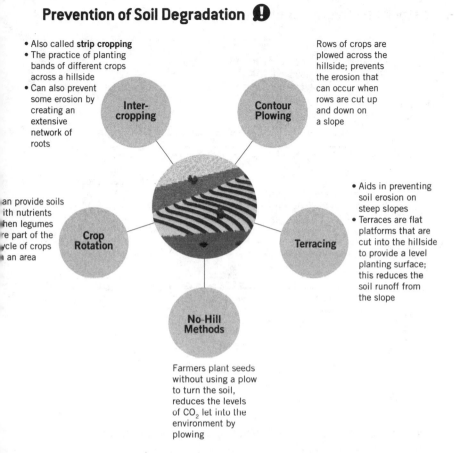

- Also called **strip cropping**
- The practice of planting bands of different crops across a hillside
- Can also prevent some erosion by creating an extensive network of roots

Inter-cropping

Rows of crops are plowed across the hillside; prevents the erosion that can occur when rows are cut up and down on a slope

Contour Plowing

...an provide soils ...ith nutrients ...hen legumes ...re part of the ...ycle of crops ...an area

Crop Rotation

Terracing

- Aids in preventing soil erosion on steep slopes
- Terraces are flat platforms that are cut into the hillside to provide a level planting surface; this reduces the soil runoff from the slope

No-Hill Methods

Farmers plant seeds without using a plow to turn the soil, reduces the levels of CO_2 let into the environment by plowing

---

Did You Know?

In the 1930s, droughts in the Great Plains reduced the area to a giant **Dust Bowl**. Although the drought was the major cause of the Dust Bowl, farming practices used at that time also contributed to the destruction of the land.

Livestock and Overgrazing ❗

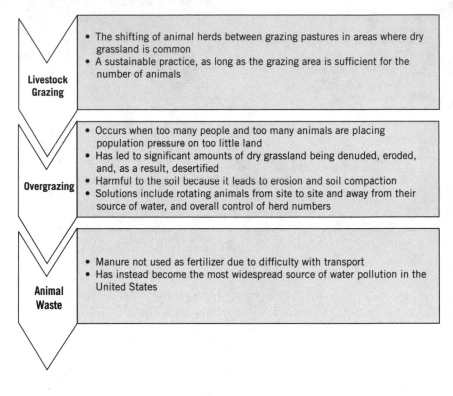

Livestock Grazing
- The shifting of animal herds between grazing pastures in areas where dry grassland is common
- A sustainable practice, as long as the grazing area is sufficient for the number of animals

Overgrazing
- Occurs when too many people and too many animals are placing population pressure on too little land
- Has led to significant amounts of dry grassland being denuded, eroded, and, as a result, desertified
- Harmful to the soil because it leads to erosion and soil compaction
- Solutions include rotating animals from site to site and away from their source of water, and overall control of herd numbers

Animal Waste
- Manure not used as fertilizer due to difficulty with transport
- Has instead become the most widespread source of water pollution in the United States

Grazing animals also consume 70% of the total grain crop consumed in the United States, making them expensive foodstuff.

 Fun fact about animal waste: one of the most expensive types of coffee is made from beans that come from the poop of a palm civet. The coffee? It's called kopi luwak. Enjoy!

Forestry !

Many environmentalists are concerned about the deforestation that is taking place in North America.

 ## *Did You Know?*

The number of trees growing in North America is approximately the same as 100 years ago, but only 5% of the original forests are left.

Deforestation !

Deforestation, or the removal of trees for agricultural purposes or purposes of exportation, is a major issue for conservationists and environmentalists. Worldwide, industrialized countries have a higher demand for wood and less deforestation, while developing countries exhibit a smaller demand for wood but more deforestation.

Nearly all of the deforestation that takes place in North America is done in order to create space for homes and agricultural plots. In sites where deforestation is occurring, the impact on resident ecosystems is significant.

A component of traditional agriculture that's still practiced in many developing countries today is a method called **slash and burn**, which severely reduces the amount of available forest and contributes to deforestation. In slash and burn, an area of vegetation is cut down and burned before being planted with crops. Then, because soils in these developing countries are generally poor, the farmer must leave the area after a relatively short time and find another location to clear.

The use of slash-and-burn farming technique in Ceará, Brazil contributes to the already mass-scale deforestation in the Amazonian forest.

Another environmentally negative by-product of deforestation is seen in countries with tropical forests. In these forests, when trees are removed and farms are placed in the cleared land, the already-poor soil is further degraded, and the area can only support crops for a short time. The negative repercussions of clearing tropical rainforests include losses in biodiversity and the erosion and depletion of nutrients in the soil.

Tree Plantations and Old Growth Forests 🛑

There are three major types of forests, which are categorized based on the age and structure of their trees: **old growth forests, second growth forests,** and **plantations** or **tree farms.**

About 95% of the world's forests are naturally occurring, and the remaining forests are known as plantations or tree farms.

Old Growth Forests	Second Growth Forests	Plantations and Tree Farms
• Have never been cut or seriously disturbed for several hundred years • Contain incredible biodiversity, with myriad habitats and highly evolved, intricate niches for a multitude of organisms	Areas where cutting has occurred and a new, younger forest has arisen naturally	Planted and managed tracts of trees of the same age that are harvested for commercial use

Old Growth Forests Over Time

Forest Management ❗

The management of forest plantations for the purpose of harvesting timber is called **silviculture**. The basic purpose of this is to create a sustainable yield; humans must harvest only as many trees as they can replace through planting. There are two basic management plans:

1. **Clear-cutting** is the removal of all of the trees in an area. This is typically done in areas that support fast-growing trees, such as pines. This is the most efficient way for humans to harvest the trees, but it has a major impact on the habitat.

2. **Selective cutting** is the removal of select trees in an area. This leaves the majority of the habitat in place and has less of an impact on the ecosystem. This type of **uneven-aged management** is more common in areas with trees that take longer to grow or if the forester is only interested in one or more specific types of trees that grow in the area.

Another type of uneven-aged management occurs in shelter-wood cutting. For **shelter-wood cutting**, mature trees are cut over a period of time (usually 10–20 years); this leaves some mature trees in place to reseed the forest.

In the case of **agroforestry**, trees and crops are planted together. This creates a mutualistic symbiotic relationship between the trees and crops: the trees create habitats for animals that prey upon the pests that harm crops, and their roots also stabilize and enrich the soil.

National Forest Policy ❗

The federal government owns about 28% of all land in the United States. The need to preserve some of the land was recognized by President Lincoln, who set aside a park in Yosemite, California. In 1916 the **National Park System** was created in part to manage and preserve forests and grasslands.

Today, in addition to the National Park System, there are several ways the federal government controls forested land:

- **Wilderness preservation areas** are open only for recreational activities with no logging permitted.
- The National Forest System, Natural Resource Lands, and National Wildlife Refuges are the other groups of federally controlled lands that allow logging with a permit.
- Two important laws that relate to our federal government's policies on preserving public lands are the **Wilderness Act** and the **Wild and Scenic Rivers Act**.

Date	Name of Legislation	What It Did
1964	Wilderness Act	Established a review of road-free areas of 5,000 acres or more and islands within the National Wildlife Refuges or the National Park System for inclusion in the National Preservation System. This act restricted activities in these areas.
1968	Wild and Scenic Rivers Act	Established a National Wild and Scenic Rivers System for the protection of rivers with important scenic, recreational, fish and wildlife, and other values.

Natural Events and Forest Fires ❗

Certain tree diseases and tree pests are natural problems in forested areas that affect the quality of the food and the number of trees that are available for use. Humans manage these natural events in many different ways:

- removing infected trees
- removing select trees or planting them sparsely to provide adequate spacing between them
- using chemical and natural pest controls
- carefully inspecting imported trees and tree products
- developing pest- and disease-resistant species of trees through genetic engineering.

Forest fires are another natural occurrence and are part of the natural life of a forest. Under natural conditions, fires burn every few years and consume the dry leaves, needles, and wood on the forest floor. However, if there are fewer fires, the amount of fuel can build up to very high levels.

One way to solve the fuel buildup issue is to implement **controlled burns**, or small fires started when the conditions are just right and which lower the amounts of fuel; this practice is quite controversial.

You should know the following types of forest fires that occur naturally:

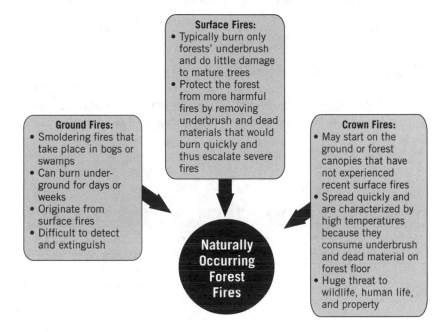

Surface Fires:
- Typically burn only forests' underbrush and do little damage to mature trees
- Protect the forest from more harmful fires by removing underbrush and dead materials that would burn quickly and thus escalate severe fires

Ground Fires:
- Smoldering fires that take place in bogs or swamps
- Can burn underground for days or weeks
- Originate from surface fires
- Difficult to detect and extinguish

Crown Fires:
- May start on the ground or forest canopies that have not experienced recent surface fires
- Spread quickly and are characterized by high temperatures because they consume underbrush and dead material on forest floor
- Huge threat to wildlife, human life, and property

Naturally Occurring Forest Fires

Rangelands

In the United States, **rangelands** are diverse in nature, consisting of desert, wet and dry grasslands, and mountainous meadows. Such diverse ecosystems allow for the grazing of animals, wildlife habitats, renewable and nonrenewable energy resources, minerals, lumber, and recreational opportunities.

There are approximately 770 million acres of rangelands in the United States, of which the federal government manages the 43%; roughly 96 million acres of the lands managed by the National Forest System are deemed rangelands.

Due to their diverse and rich nature, rangelands are used by farmers and livestock owners for food production and grazing lands. As populations have increased, the sustainability of such rangelands has been compromised: too many people and too many animals are placing population pressure on too little land. As discussed in Chapter 3, human activity has led to overgrazing, deforestation, and soil salinization; these activities can all lead to desertification.

Desertification is any human process that turns a vegetated environment into a desert-like landscape.

Rangeland Management ❗

Various tracts of public lands are available for use as rangeland, and cooperation between government agents, environmentalists, and ranchers can help avoid problems of overgrazing on these lands.

The **Bureau of Land Management** is responsible for managing federal rangelands. The following legislation was passed in order to regulate the use of rangelands throughout the United States.

Date	Name of Legislation	What It Did
1934	Taylor Grazing Act	The first federal effort to regulate grazing on federal public lands. The act established grazing districts and enacted a permit system in order to improve rangeland conditions through regulated use of rangelands for livestock grazing.
1936	Soil Conservation and Domestic Allotment Act	A follow-up to the Soil Conservation Act of 1935, this act was intended to encourage soil usage that preserved and improved soil fertility, promoted economic use, and reduced the exploitation and fiscally irresponsible use of soil resources. In an attempt to conserve soil and prevent erosion, this act permitted the government to pay farmers.

Mining

Mining is the excavation of earth for the purpose of extracting ore or minerals. Mineral resources are divided into categories depending on how they're used.

Mineral Resource	What Is It?
Metallic minerals	Mined for their metals such as zinc, which can be extracted through smelting and used for various purposes
Nonmetallic minerals	Mined for their use in their natural state—nothing is extracted from them; e.g., salt and precious gems
Mineral deposit	Area in which a particular mineral is concentrated
Ore	A rock or mineral from which a valuable substance can be extracted at a profit

Mineral Formation

The oldest rocks on Earth are 3.8 billion years old, while others are only a few million years old. This means that rocks have to be recycled. The process that does this is the **rock cycle.**

In the rock cycle, time, pressure, and the Earth's heat interact to create three basic types of rocks: **igneous, sedimentary,** and **metamorphic.** Refer back to Chapter 1 to review the rock cycle and the characteristics of each type of rock.

Mineral Extraction ❗

The cost of extracting minerals depends on numerous factors, including the location and size of the mineral deposit. Mining can lead to the following environmental issues:

- Ecosystem harm and destruction
- Pollutants associated with the mineral ore underground or machinery used
- **Acid mine drainage**, resulting from coal mining, harms local stream ecosystems
- Excess waste material called **gangue** and piles of gangues called **tailings**
- Extensive energy input to refine minerals post-extraction

Types of Mining ❗

Strip Mining ❗

A controversial type of mining is **strip mining**, which involves stripping the surface layer of soil and rock, or **overburden**, in order to expose a seam of mineral ore. Both the least expensive and least dangerous method of mining for coal, strip mining is only practical when the ore is relatively close to the surface. Unfortunately, strip mining has a more detrimental effect on the environment than does underground mining.

Mountaintop Removal ❗

The most extreme form of strip mining, **mountaintop removal**, which is mostly associated with coal mining in the Appalachian Mountains, transforms the summits of mountains and destroys ecosystems.

Shaft Mining ❗

With **shaft mining**, vertical tunnels are built to access and then excavate minerals that are underground and otherwise unreachable.

ASAP Environmental Science

Mine Restoration ⚠️

Fortunately, air, land, and water harmed by mining can be reclaimed through **mine restoration** projects. In 1977, Congress passed the **Surface Mining Control and Reclamation Act** (SMCRA), which created programs to help coal mines manage pollutants and to guide the reclamation of abandoned mines.

Global Reserves ⚠️

The following tables show the production (in thousands of metric tons) of some non-fuel mineral resources. You do not need to memorize these figures for the AP exam, but remember that these high production rates lead to the eventual depletion of these resources. Also, be ready to describe the impact of mining and mineral production on the environment.

Mineral	2016 Production (metric tons)
Copper	19,400
Phosphate rock (for fertilizer)	261,000
Bauxite (aluminum ore)	262,000
Iron ore (usable)	2,230,000
Zinc	11,900,000

Mineral	2016 Reserves (metric tons)
Bauxite (aluminum ore)	28,000,000
Copper	720,000
Iron ore (crude ore))	170,000
Nickel	78,000,000
Zinc	220,000

As you can see, demand for mineral resources is very high. The increased demand for manufactured goods indicates that we need to extract more and more raw materials from the earth. For the exam, you should be aware of the need to use mineral resources in a sustainable manner.

Relevant Laws and Treaties ❗

There are several laws that govern mining in the U.S. and address the exploration and mining of minerals, as well as the negative impact of waste and pollution that result from mining.

Date	Name of Legislation	What It Did
1872	Mining Act	Governed prospecting and mining of minerals on publicly owned land
1920	Mineral Leasing Act	Permitted the Bureau of Land Management to grant leases for development of deposits of coal, phosphate, potash, sodium, sulphur, and other leasable minerals on public domain lands
1976	Resource Conservation and Recovery Acts (RCRA)	Regulated some mineral processing wastes
1977	Surface Mining Control And Reclamation Act	Established a program for regulating surface coal mining and reclamation activities; established mandatory standards for these activities on state and federal lands, including a requirement that adverse impacts on fish, wildlife, and related environmental values be minimized
1980	Comprehensive Environmental Response, Compensation, and Liability Act (Superfund)	Regulated damage done by mining

Fishing

The term **fishery** is used in several ways, but it is primarily defined as the industry or occupation devoted to the catching, processing, or selling of fish, shellfish, or other aquatic animals.

Fishing Techniques

Most of the fish that are harvested worldwide come from **capture fisheries**; they are caught in the wild and not raised in captivity for consumption. Some of the techniques that have been developed in order to improve fishing yields are creating problems that relate to overfishing.

One of these problems is known as by-catch. **By-catch** refers to any other species of fish, mammals, or birds that are caught that are not the target fish. Some fishing methods that result in by-catch are

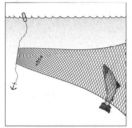
Drift Nets
Drift nets float through the water and catch everything in their path.

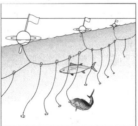
Long-Lining
Long-lining is the use of long lines with baited hooks that are taken by numerous aquatic organisms.

Bottom-Trawling
In bottom-trawling, the ocean floor is literally scraped by heavy nets that smash everything in their path.

Some advances that have been made in the fishing industry in an attempt to mitigate the problems of by-catch are restrictions on the use of drift nets, the installation of ribbons on bait hooks that scare away birds and prevent them from being caught, and bans on bottom trawling.

Overfishing and Aquaculture 〰

Recent reports show that

- About 69% of the major fish stocks of the world are either overexploited or fully exploited.
- About 20% of the stocks are moderately overexploited.
- About 8% are either recovering from depletion or depleted.

This is mostly due to overfishing.

One partial solution to the problem of overfishing is **aquaculture**, or **fish farming**, which is the raising of fish and other aquatic species in captivity for harvest. In general, the fish that are raised in captivity are those with the highest economic value.

Different methods are used in aquaculture—some fish are raised totally in captivity and then harvested, while others are hatched in captivity before release into the wild. Some **saltwater aquaculture** occurs in shallow coastal areas, though this is generally for raising sea plants and mollusks.

Relevant Laws and Treaties ❗

You should know the following U.S. laws pertaining to preserving ocean resources.

Date	Name of Legislation	What It Did
1965	Anadromous Fish Conservation Act	Protected fish that live in the sea but grow up and breed in freshwater
1972	Marine Mammal Protection Act	Established a federal responsibility to conserve marine mammals
1973	Endangered Species Act	Provided broad protection for species of fish, wildlife, and plants that are listed as threatened or endangered in the U.S. or elsewhere
1976	Magnuson Fishery Conservation and Management Act	Governed the conservation and management of ocean fishing

You should also be familiar with the following international agreements:

Date	Name of Legislation	What It Did
1975	CITES (the Convention on International Trade in Endangered Species of Wild Fauna and Flora)	An international agreement between governments that ensured that international trade in specimens of wild animals and plants do not threaten their survival
1982	The United Nations Agreement for the Implementation of the Provisions of the United Nations Convention on the Law of the Sea	Set out the principles for the conservation and management of certain types of fish

Chapter 4 Summary

- The "Tragedy of the Commons" is essential to understanding the management of shared resources.
- Renewable resources are perpetual or replenished quickly (~50 years or less); nonrenewable resources renew at an insufficient rate for human use.
- There are two main methods of agriculture: traditional subsistence and industrial.
- Deforestation is the largest concern in forestry management, and several policies have been enacted to protect the age and diversity of forests.
- Forest fires can be essential to plant regeneration and forest health but can also burn uncontrolled and cause great economic loss.
- Mining: Both naturally used and metallic minerals have great value and functions, but they also have a significant environmental impact due to extraction pollution, soil erosion, and habitat loss.
- Improved technologies, fishing strategies, and human population growth have contributed to overfishing (harvesting fish faster than the population can replenish themselves).

CHAPTER 5
Energy Resources and Consumption

As populations increase, more pressure is placed on Earth's natural resources, and along with this comes the need for humans to find ways to develop those natural resources for direct human use. In this chapter, we will review energy resources and consumption.

Energy Concepts

Unlike the essential elements we discussed in earlier chapters, energy flows on a one-way path through the atmosphere, hydrosphere, and biosphere. It is essential for living organisms in many of its forms.

Forms of Energy

Energy is the capacity to do work. There are several types of energy:

- **Potential:** energy at rest, or stored energy
- **Kinetic:** energy in motion
- **Radiant:** sunlight, or electromagnetic energy
- **Thermal:** heat, or the internal energy in substances
- **Chemical:** energy stored in chemical bonds between atoms
- **Electrical:** energy from the motion of electrons
- **Nuclear:** stored in the nuclei of atoms

 Remember!

Potential energy can be converted to kinetic energy. You may remember this from physics class.

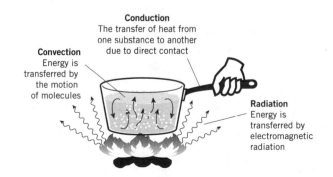 *Did You Know?*

Radiant energy is the only form of energy that can travel through empty space.

Energy can move around Earth in three ways:

Conduction
The transfer of heat from one substance to another due to direct contact

Convection
Energy is transferred by the motion of molecules

Radiation
Energy is transferred by electromagnetic radiation

Convection and conduction are very important processes that drive the movement of water in the hydrosphere of Earth and affect our weather.

Units of Energy 🛈

You should recognize the following units of energy and power:

Energy	• Joule (J) • Calorie (cal) • British thermal unit (BTU) • Kilowatt hour (kWh), a measure of watt/time
Power	• Watt (W) • Horsepower (hp)

Note that a watt is equal to *volts* × *amperage.*

Converting Between Units 🛈

- There are two systems of units you should know:
  - SI: International System
  - USCS: United States Customary System
- All unit systems are based on a combination of fundamental and derived units.
- All derived units come from fundamental units, using fundamental physical laws.
- Here are the fundamental units for time, length, and mass:

Quantity	SI Unit	USCS Unit
Time	Second (s)	Second (s)
Length	Meter (m)	Foot (ft)
Mass	Kilogram (kg)	Pound mass (lbm)

Did You Know?

Work 🔔

Force is calculated as *mass × acceleration*. Here are some commonly used units:

Force =	Mass ×	Acceleration
Newton	kg	m/s^2
dyne	gram	cm/s^2
lb	slug	ft/s^2

Energy 🔔

Force × distance gives you a numerical value for energy. Here are some commonly used units:

Energy =	Force ×	Distance
joule	newton	meter
erg	dyne	cm
ft-lb	lb	ft

Laws of Thermodynamics ❗

Law	Definition	Example 〜
First	• Energy can neither be created nor destroyed. • It can only be transferred and transformed.	In photosynthesis, radiant energy from the Sun is converted to chemical energy in the form of the bonds that hold together atoms in carbohydrates.
Second	• Entropy (disorder) of the universe is increasing. • In most energy transformations, a significant fraction of energy is lost to the universe as heat.	In food chains, only about 10% of the energy from one trophic level is available for the next level upon consumption.

Sources of Energy ❗

The rest of this chapter will focus on reviewing different sources of energy. The Earth provides us with energy resources just as it provides the physical, natural resources we learned about in the last chapter.

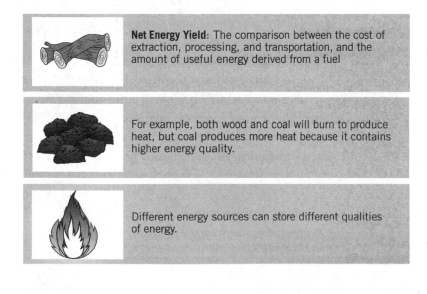

Net Energy Yield: The comparison between the cost of extraction, processing, and transportation, and the amount of useful energy derived from a fuel

For example, both wood and coal will burn to produce heat, but coal produces more heat because it contains higher energy quality.

Different energy sources can store different qualities of energy.

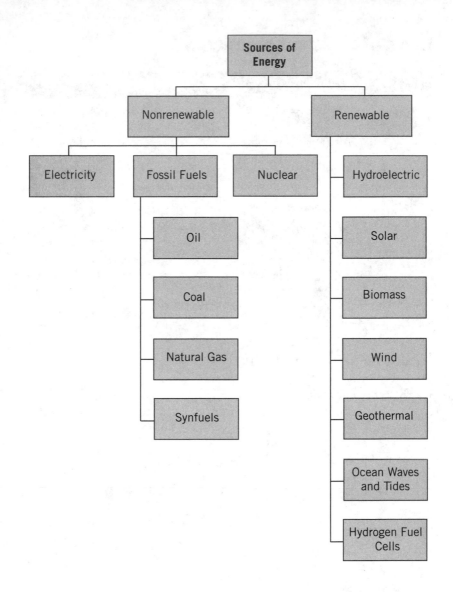

Electricity and Energy Consumption ❗

Producing electricity is one of the biggest uses of energy. In other words, each year we use tons of energy to produce another form of energy (electricity).

How Is Electricity Made? 〰️

Step 1. An energy source provides power to heat up water.
Step 2. Water is transformed into steam (kinetic energy).
Step 3. Steam turns a turbine (mechanical energy).
Step 4. Magnets in a generator pass over copper wire coils (or vice versa).
Step 5. A flow of electrons through the copper wire is generated.
Step 6. This produces an alternating current that passes into electrical transmission lines.

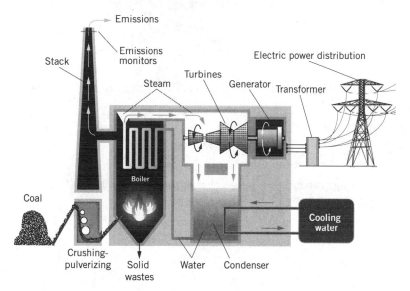

Emissions

Emissions monitors

Stack

Steam

Turbines

Generator

Transformer

Electric power distribution

Boiler

Coal

Crushing-pulverizing

Solid wastes

Water

Condenser

Cooling water

Did You Know?

- In lieu of steam, flowing water or wind can also provide the power needed to turn the turbine and produce electricity.
- A generator consists of copper wire coils and magnets. One is stationary (the stator) and the other rotates (the rotor).

In the first step of creating electricity, energy is used to heat up water. This energy comes from:

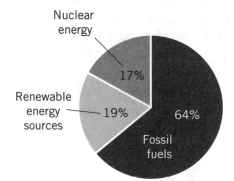

These numbers represent global energy sources used to power the world's electricity.

Fossil Fuels !

The fossil fuels are oil, coal, and natural gas. They provide 80 percent of the world's energy and are a nonrenewable energy source.

What Are Fossil Fuels?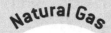

Oil

Long chains of hydrocarbons

Coal

Mixture of carbon, hydrogen, oxygen, and other atoms

Natural Gas

Mostly methane gas (CH_4) with a mixture of other gases (such as pentane or butane) in small quantities

---

 Did You Know?

During the Industrial Revolution in the early 18th century, mechanical processes were powered by burning firewood and coal to produce steam.

Note that for the sake of the AP exam, oil and petroleum are interchangeable terms.

There are different types of coal, ranked by the number of BTUs they produce upon burning.

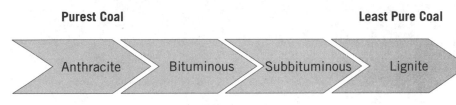

Purest Coal   **Least Pure Coal**

Anthracite   Bituminous   Subbituminous   Lignite

Where Are Fossil Fuels Found? !

Fossil fuels are formed from the fossilized remains of once-living organisms. Over long periods of time, organic matter was exposed to intense heat and pressure. Eventually, these forces broke the organic molecules down into oil, coal, and natural gas.

Fossil Fuel	Source	Locations
Oil and Natural Gas	Ancient marine organisms such as zooplankton	• Deep in the earth • Under both land and ocean floor • Stored in spaces between rocks
Coal	• Organisms in ancient swamps, especially plants	• Long continuous deposits called seams • At various depths underground
	• Living organisms (mostly anaerobic bacteria) found in landfills, swamps, wetlands, and the intestines of various animals.	• Many locations around the world

Tar sands are the dirtiest of all the oils extracted. This method of extraction and processing are detrimental to the environment and use a great deal of energy.

Did You Know?

- The largest source of methane is wetlands, and the second largest source is flatulent livestock!
- Fossil fuels are still forming today, but they will not be available for use for many generations.

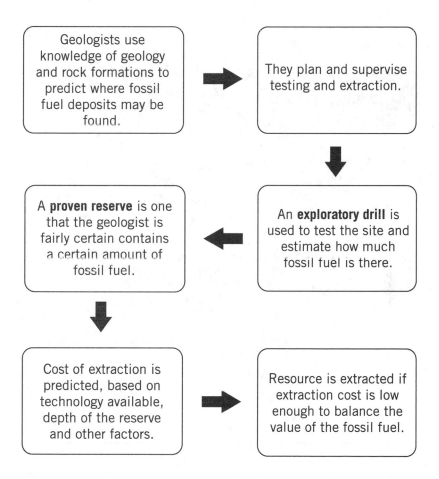

Geologists use knowledge of geology and rock formations to predict where fossil fuel deposits may be found.

➡ They plan and supervise testing and extraction.

⬇

An **exploratory drill** is used to test the site and estimate how much fossil fuel is there.

⬅ A **proven reserve** is one that the geologist is fairly certain contains a certain amount of fossil fuel.

⬇

Cost of extraction is predicted, based on technology available, depth of the reserve and other factors.

➡ Resource is extracted if extraction cost is low enough to balance the value of the fossil fuel.

Extraction and Use of Fossil Fuels 🛑

Crude Oil 🛑

Crude oil is the name of raw or fresh oil when it is pumped up from a reserve. Crude oil varies greatly from reserve to reserve. It can be thin or viscous (thick), have a range of sulfur content, and vary in color and odor. Crude oil is used for the production of fuel products, plastics, and petroleum jelly.

There are three different methods of extracting oil:

Primary Extraction	• Oil well is tapped and pumped to the surface. • This is the easiest way to extract oil.
Pressure Extraction	• This method is used when oil is harder to extract. • Mud, saltwater, or CO_2 is used to push oil out from the reserve.
Heat Extraction	• Steam, hot water, or hot gases are used to partially melt very thick crude oil. • This makes the oil easier to extract.

 Did You Know?

When some oil wells are tapped for the first time, there is a large release of oil and gas due to the pressure in the reserve. This is called a gusher.

Coal ❗

Coal mining occurs through one of two processes, both of which can be hazardous and have serious environmental impacts.

<table>
<tr><th>Strip Mining</th><th>Underground Mining</th></tr>
<tr><td>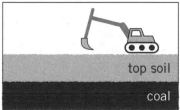</td><td>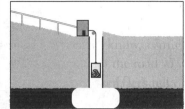</td></tr>
</table>

Strip Mining	Underground Mining
• Removal of the earth's surface, all the way down to the level of the coal seam • Removed earth is called the **overburden** • Coal is removed; overburden is replaced with and topped with soil • Area contoured and re-vegetated	• Sink shafts to reach underground deposits • Networks of tunnels are dug or blasted • Humans enter these tunnels to manually retrieve coal

top soil
coal

Con: *Leaves behind a slag (man-made mound of waste material as a by-product of coal mining), which is hazardous. Sulfur in the slag can leach out and enter the water table.*

Con: *After production stops at underground mines, cave-ins can occur, causing massive slumping or subsidence.*

At this time, coal is one of the two most abundant fossil fuels and is used to generate electricity in 30% of the power plants in the United States.

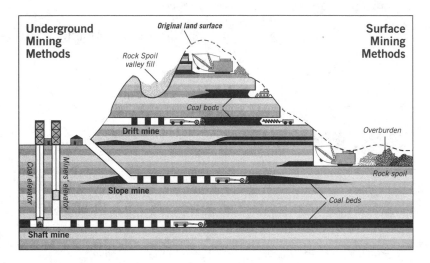

Natural Gas 😮

Uses	• Heating homes • Cooking • Burned to generate electricity • Can be used in some car and truck engines
Pros	• Produces only CO_2 and H_2O when it burns • Cleaner than burning oil or coal
Cons	• A leak can cause violent explosions • More difficult to transport than coal or oil • Liquefied before transportation, which requires energy • Can be transported via pipes, but these can leak and explode, as well as damage habitats when they're installed

Did You Know?

- Some power plants are designed to switch between oil and natural gas fuels, depending on cost fluctuations.
- In the future, we may be able to efficiently trap methane being released from landfills.

How Much Fossil Fuel Is Left? ❗

Global Demand ❗

In order to understand how long our accessible fossil fuel supplies will last, you should know how quickly we are using up those fuels. Here is some data from 2015, the last year for which the U.S. Energy Information Administration has this information:

Country	2015 Petroleum Consumption (thousands of barrels per day)
United States	19,533
China	12,376
Japan	4,142
India	4,120
Russia	3,512

 Did You Know?

The five countries in the table above alone consume almost 39.2 million barrels of oil each day!

M. King Hubbert 💬

M. King Hubbert was an expert on energy demands.

- Geophysicist at the Shell lab in Texas
- Devised a theory about the future of oil production in the 1950s

Hubbert Peak Theory

- Theory that the end of oil as a cheap and easily available form of energy is in the near future and that we must begin to develop alternative fuel sources
- Based on the observation that the amount of oil under the ground in any region is finite
- Some parts of his theory are still used today when estimating oil production, but other parts have been disproved, such as the prediction that oil production would peak in the 1970s and then decline.

Synfuels 💬

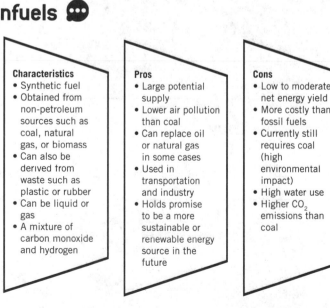

Characteristics
- Synthetic fuel
- Obtained from non-petroleum sources such as coal, natural gas, or biomass
- Can also be derived from waste such as plastic or rubber
- Can be liquid or gas
- A mixture of carbon monoxide and hydrogen

Pros
- Large potential supply
- Lower air pollution than coal
- Can replace oil or natural gas in some cases
- Used in transportation and industry
- Holds promise to be a more sustainable or renewable energy source in the future

Cons
- Low to moderate net energy yield
- More costly than fossil fuels
- Currently still requires coal (high environmental impact)
- High water use
- Higher CO_2 emissions than coal

ASAP Environmental Science

Environmental Considerations of Fossil Fuels ⚠️

Oil Drilling ⚠️

Pros
- Drilling for oil is only moderately damaging to the environment.
- Little land is needed to drill.

Cons
- Oil is transported thousands of miles by tankers, pipelines, and trucks.
- A lot of environmental damage can occur during transportation.

 Did You Know?

- In 2010, there was an explosion on the **Deepwater Horizon** drilling rig. Eleven men were killed, and oil spilled into the Gulf of Mexico for three months. This was the largest marine oil spill in the history of the oil industry. It caused a great deal of damage to marine and wildlife habitats, as well as the economy.
- In 1989, an oil tanker called the **Exxon Valdez** struck a reef in Alaska, spilling 10.8 million gallons of crude oil over several days. This is considered one of the worst environmental disasters caused by humans, and is the second largest oil spill in U.S. waters.

Burning Coal ⚠️

Burning coal produces:

- air pollutants such as carbon dioxide, nitrogen oxides, mercury, and sulfur dioxide
- fly ash—small dark flecks that are carried into the air
- boiler residue—solid waste left at the bottom of the boiler

Coal also contains iron sulfide (pyrite), which can be removed by grinding the coal into small lumps and washing it.

💬 Smokestack scrubbers remove air pollutants (especially sulfur dioxide) from burning coal emissions. For example, scrubbers often contain alkaline substances that precipitate out sulfur dioxide. This forms calcium sulfate, which is eliminated in waste sludge.

There are different types of smokestack scrubbing:

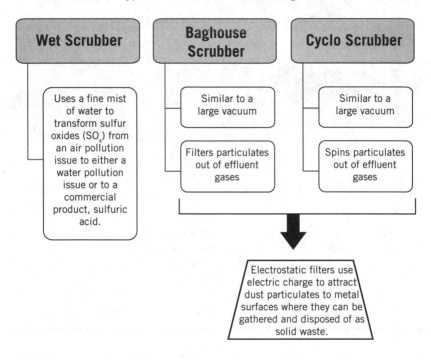

Did You Know?

- Wet scrubbing is similar to how a rainstorm will cleanse the air of pollen and dust that causes allergies in some people.
- The electrostatic filters in baghouse scrubbers and cyclo scrubbers are similar to your television screen being the dustiest surface in your living room due to electric properties of the device.
- Electrostatic precipitators help to remove 98% of particulate matter from fuel emissions.

Controversies Surrounding Fossil Fuels 🔊

Fracking

- Fracking is hydraulic fracturing—a process to extract natural gas and oil from rock that lies deep underground.
- A deep well is drilled, and millions of gallons of fracking fluid are injected at high pressure; this fractures the rock and releases oil or gas.
- Fracking is a highly controversial process. It is linked to earthquakes in Arkansas, Ohio, and Pennsylvania. It also requires a large amount of water, which ends up highly contaminated and needs to be safely stored.

Keystone Pipeline

- An oil pipeline from Alberta, Canada, south to the refineries in the Midwest and along the Gulf Coast
- An additional branch of the pipeline has been proposed but has not been completed due to concerns over pollution, such as contaminated groundwater.

Mercury

- Coal-fired power plants are the major source of mercury pollution in the environment.
- Airborne mercury can travel hundreds of miles.
- Mercury contaminates lakes, streams, and other bodies of water. It also accumulates in fish, which are then eaten by people.

Abandoned Metal and Coal Mines

- Frequently produce acid mine drainage, highly acidic water that flows to surrounding areas

Did You Know?

An EPA study found that one in six women of childbearing age in the U.S. may have blood mercury levels that could be harmful to a developing fetus.

Nuclear Energy

Nuclear energy is the world's primary non-fossil fuel, nonrenewable energy source. In the United States, 20% of electrical energy is provided by nuclear power plants. Worldwide, more than 400 nuclear power plants produce approximately 13% of the world's electrical energy.

The Big Picture: Nuclear power plants harvest energy from nuclear reactions that release nuclear energy, and convert it to electricity.

Fission and Fusion

There are two fundamental nuclear processes considered for energy production: **fusion** and **fission**.

Fusion	Fission
• Combining two small atoms (such as hydrogen or helium) to produce heavier atoms and energy • Can release more energy than fission without producing as many radioactive by-products • Reactions occur in the Sun • This type of technology is still being developed, but holds promise for the future of nuclear energy.	• Discovered first • Energetic splitting of large atoms (such as uranium or plutonium) into two smaller atoms (called fission products) • To split an atom, you have to hit it with a neutron. • Several neutrons are released, which continue the chain reaction. • Used by all commercial nuclear power plants in operation

A common fusion example involves two isotopes of hydrogen: tritium (which contains 2 neutrons) and deuterium (which contains 1 neutron).

Current nuclear plants use uranium-238 enriched with 3% uranium-235. ^{235}U splits via fission, initiating the nuclear reaction.

Breeder reactors generate new fissionable material faster than they consume such material. They often use ^{238}U, plutonium-239 (^{239}Pu), or thorium-234 (^{234}Th).

Radioactive materials all have half-lives, which is the time it takes for half of the radioactive sample to degrade.

The future of nuclear power will possibly involve nuclear fusion, though this area of research and technology development has been quite difficult and progress has been much slower than expected.

In the U.S., there are two types of nuclear reactors: boiling water reactors and pressurized water reactors.

---

^{238}U and ^{235}U are isotopes of uranium: they have the same number of protons, but a different number of neutrons (and thus different masses).

Nuclear Reactors 🔘

Boiling Water Reactor 🔘

- This involves using heat of the reactor core to boil water into steam.
- Steam is piped directly to the turbines.
- Steam spins turbines to generate electricity.
- Water is cooled back to a liquid (by a heat exchanger), and then pumped back to the core to be turned into steam again.
- This process uses two water circulation systems: one makes steam and carries it to the turbine, and the other cools the water from the core so it can be turned back into steam.

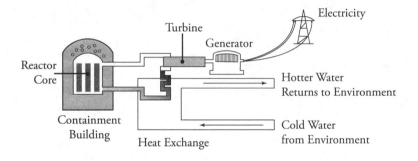

Pressurized Water Reactor ⚠

- A pressurized water reactor produces electricity by generating steam.
- This reactor contains three water circulation systems:
  1. Kept under high pressure to prevent the water from boiling; it passes through the reactor heat exchanger and transfers its energy to the second water supply
  2. Not kept under high pressure, forms steam to spin the turbines
  3. Cools steam from the turbines to regenerate the second water supply

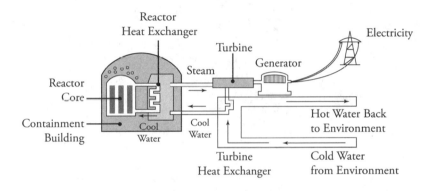

Advantages and Disadvantages of Nuclear Power ⚠

Benefits ⚠

Nuclear power plants have low operating costs and virtually none of the emissions that lead to smog, acid rain, or global warming.

 Remember!

Production and combustion of fossil fuels produces harmful gases such as sulfur dioxide, nitrogen oxide, and carbon dioxide.

Nuclear Energy Safety Issues ❶

There are a number of safety issues associated with nuclear energy.

Nuclear Energy Safety Issues	
Safety Issue	Description
Meltdown	Reactor loses coolant water, and thus the very hot core melts through the containment building. The radioactive materials could then get into the groundwater.
Explosion	Gases generated by an uncontrolled core burst the containment vessel and spread radioactive materials in the environment.
Nuclear weapons	Some of the by-products of the fission reaction can be remade into fission bombs, or "dirty bombs," that spread damaging radioactive isotopes.
Highly radioactive waste	No longer usable cores, piping, and spent **fuel rods** need to be stored for many centuries. The "spent" fuel can contain radioactive elements like plutonium-239 that has a half-life of 2.13×10^6 years.
Thermal pollution	The water used to cool turbines is returned to local bodies of water at a much higher temperature than when it was removed unless first cooled.
Radioactive elements	Gamma rays produced by radioactive decay can damage cells and DNA, which can cause leukemia as well as breast, thyroid, and stomach cancer. Damage to the immune system can also result.
Concern for one's safety	People suffer from mental stress, anxiety, and depression caused by concerns for their safety, resulting in **Not In My Backyard** syndrome **(NIMBY)**.

ASAP Environmental Science

Chernobyl facility in the Ukraine in 1986; devastating explosion

Three Mile Island facility in Pennsylvania in 1979; nuclear reactor accident

Fukushima nuclear power plant in Japan in 2011; nuclear reactor accident

**Famous Nuclear Reactor
Incidents**

Future of Nuclear Power 😵

China and India lead the world in the creation of new nuclear facilities. The United States is closing nuclear power plants instead of building more due to factors including

- cost for building a new nuclear power plant is prohibitive due to changing regulations.
- the need for a better long-term plan for storing nuclear waste that's created.
- insurance companies will not insure nuclear power plants, which leaves this important issue to the federal government.

Renewable Energy

There are seven types of renewable energy sources you need to know about for the AP Environmental Science Exam:

1. Hydroelectric
2. Solar
3. Biomass
4. Wind
5. Geothermal
6. Ocean waves and tides
7. Hydrogen fuel cells

Currently, renewable energy sources account for less than 15% of global energy use. The following table outlines each of these renewable energy sources, though we will review them in more detail throughout this chapter.

Did You Know?

Renewable does not always mean "clean" and *nonrenewable* does not always mean "dirty"; these terms refer to how long it takes to regenerate the fuel supply (nonrenewable is generally more than 50 years or a human's life span).

Energy Source	Definition/ Examples	Pros	Cons
Hydroelectric	• Electricity generated as moving water turns a turbine	• Doesn't produce chemical pollutants	• Produces thermal pollution • Affects river flow and nearby habitats • Limited by the number of rivers with sufficient flow and drop
Solar	• Solar panels convert solar energy into heat or electricity	• Doesn't produce pollutants • No moving parts • Require little maintenance • Silent • Pays off in the long term	• Producing PV* cells requires fossil fuels • Not every location receives enough light to make solar panels worthwhile • Initially expensive (some states give financial assistance) • Limited by Sun exposure and energy storage potential
Biomass	• Wood • Charcoal • Animal waste products	• Carbon neutral • Widely available • Still being optimized and developed	• Expensive • Can lead to deforestation
Wind	• Wind is used to rotate blades • This causes machinery inside to rotate (mechanical energy)	• Fastest growing alternative energy source • Safe • Doesn't produce pollutants	• Expensive • Loud • Unattractive to some • Need an alternative for when there is no wind

*PV is short for **photovoltaic,** which gets its name from the process of converting light (photons) to electricity (voltage).

Energy Resources and Consumption

Energy Source	Definition/ Examples	Pros	Cons
Geothermal	• Energy that's obtained from within the Earth • Harness Earth's internal heat	• Cost effective • Reliable • Doesn't produce pollutants during typical operation • Minimal harm to land and habitats • Virtually infinite supply	• Only possible in some locations • Affects and is limited by groundwater • Salts dissolved in the water corrode machinery • Gases (CH_4, CO_2, hydrogen sulfide, NH_3) or trace toxic chemicals trapped in the water may be released as the water is used.
Ocean Waves and Tides	• Dams are erected across outlets of tidal basins. • Incoming and/or outgoing tides pass through the dam, turning turbines and generating electricity.	• Doesn't produce pollutants • Predictable • Long life span	• Expensive • Limited locations • Environmental impact still being determined
Hydrogen Fuel Cells	• Hydrogen is obtained from water or organic matter, stored, and used to generate electricity.	• Use is clean and safe (only waste is steam)	• Expensive • Making the fuel cell can release pollutants (but heat can be used to warm homes or water).

 ***Did You Know?***

Charcoal is wood that has been baked to remove water and impurities.

Hydroelectric Energy

There are some famous hydroelectric dams you should be familiar with:

Three Gorges Dam

- In China
- Largest dam in the world

Grand Coulee

- Columbia River in Washington State
- Largest dam in the U.S.

Hoover Dam

- Colorado River on border of Nevada and Arizona
- One of the most famous dams in the world

Hydroelectric Dam

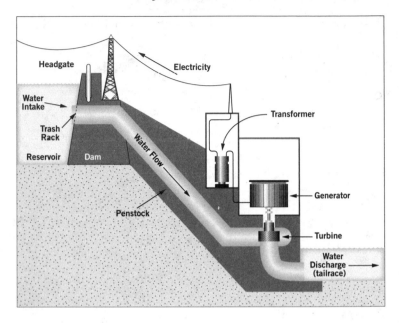

Effects of Damming a River ❗

- Change river flow rate
- Can lead to the destruction of habitats (but the creation of new ones)
- Fertilizing sediment is not passed downriver
- Increased surface area in the reservoir leads to a high rate of evaporation and water loss
- Creation of the reservoir destroys habitats: vegetation drowns and thus decomposes, releasing CO_2
- Decreased populations of salmon and other anadromous fish (dams can prevent migration between feeding and breeding zones)

Here are a few additional facts about hydroelectric dams:

Small Hydro

Dams must be strong enough to hold back many tons of sediment.

- Development of hydroelectric power on a small scale
- Suitable for local community and industry
- Common in isolated areas where a national electricity grid is not an option

A river generally has many dams along its length. The addition of each new dam leads to less and less water downstream and changes the natural course of the river.

Flood Control

- Many dams aid in flood protection and control.
- Reservoir levels are kept below a certain elevation before the onset of the rainy season.

Solar Energy !

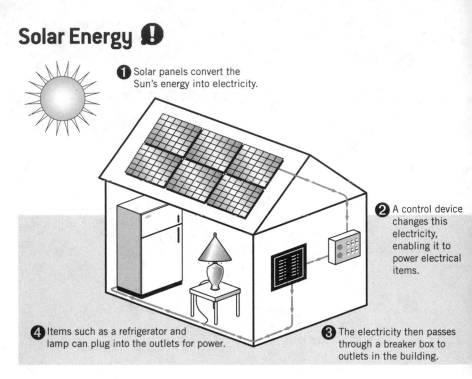

① Solar panels convert the Sun's energy into electricity.

② A control device changes this electricity, enabling it to power electrical items.

④ Items such as a refrigerator and lamp can plug into the outlets for power.

③ The electricity then passes through a breaker box to outlets in the building.

How Solar Energy Works

 Remember!
We already obtain energy we need to live through the Sun: producers capture the Sun's energy and convert it into chemical energy.

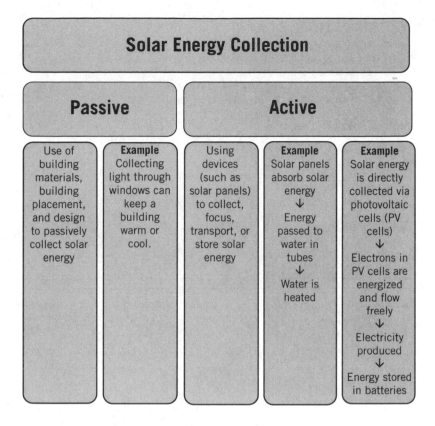

Solar Energy Collection

Passive

Use of building materials, building placement, and design to passively collect solar energy	**Example** Collecting light through windows can keep a building warm or cool.

Active

Using devices (such as solar panels) to collect, focus, transport, or store solar energy	**Example** Solar panels absorb solar energy ↓ Energy passed to water in tubes ↓ Water is heated	**Example** Solar energy is directly collected via photovoltaic cells (PV cells) ↓ Electrons in PV cells are energized and flow freely ↓ Electricity produced ↓ Energy stored in batteries

Did You Know?

Using solar energy isn't new: the Romans developed window glass, which allowed sunlight to come in and trapped solar heat indoors. The Swiss scientist Horace De Saussure built a solar reflector in 1767 that could heat water and cook food.

Homeowners with solar panels often harvest more energy than they use. This can be turned into a credit with the electricity provider or stored in batteries to be used later.

Biomass Energy

Biomass energy is renewable only when it is used at a pace that allows time for replacement.

Example: Gasohol

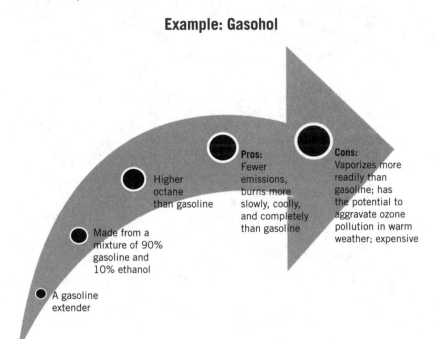

A gasoline extender

Made from a mixture of 90% gasoline and 10% ethanol

Higher octane than gasoline

Pros: Fewer emissions, burns more slowly, coolly, and completely than gasoline

Cons: Vaporizes more readily than gasoline; has the potential to aggravate ozone pollution in warm weather; expensive

Did You Know?

Ethanol is often obtained by fermenting agricultural crops or crop wastes.

Example: Biodiesel

Biodiesel is made primarily from virgin oils (such as soybean oil).

Biodiesel can also be made from waste vegetable oil.

Restaurants can use their waste oil by-products to produce biodiesel.

Algae can be grown in sewage water (without using land normally used for food production) and then used to produce biodiesel.

Wind Energy ❗

How Do Windmills Work?

Traditional Dutch Windmill

1. Wind turns the blades or paddles of the windmill.
2. This drives a shaft that's connected to several cogs.
3. The cogs turn wheels that can perform mechanical work (grinding grain or pumping water).

Modern Windmill

1. Wind blows into the wind turbine.
2. The blades spin.
3. Machinery inside the base of the windmill rotates.
4. This rotation is used to generate electricity.

Nacelle—Base of the windmill that houses a gearbox, generator, and machinery that controls the turbine

Did You Know?

Wind turbines kill between 214,000 and 368,000 birds annually.

Here are a few additional facts about wind energy:

People have been using wind to produce energy for centuries (as early as the 17th century).

Modern wind turbines are usually placed in groups called wind farms or parks.

Wind-generated power has been increasing at a rate of more than 30% per year.

Wind energy is projected to supply 20% of the world's energy needs by 2030.

American wind farms are currently predominately in California and Texas.

In the future, wind farms may also be located offshore in the ocean.

Geothermal Energy 🔴

- The interior of the Earth is still warm due to radioactive decay.
- In other words, geothermal energy indirectly gains its energy from nuclear power.
- High temperatures in the Earth result in pressure buildup.
- Some of this heat escapes through fissures and cracks to the surface, leading to geysers, hydrothermal vents, and hot springs.

How Does Geothermal Energy Production Work? 🔴

Step 1. Wells are drilled down into the Earth as far as thousands of meters to water that is 300–700°F.

Step 2. Naturally heated water and steam from the Earth's interior are brought to the surface.

Step 3. Steam powers a turbine.

Step 4. Heated water piped directly though buildings can heat them.

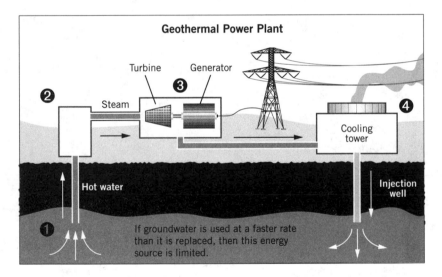

Geothermal Power Plant

Ocean Tides 💬

Tidal movements of ocean water can be tapped and used as a source of energy.

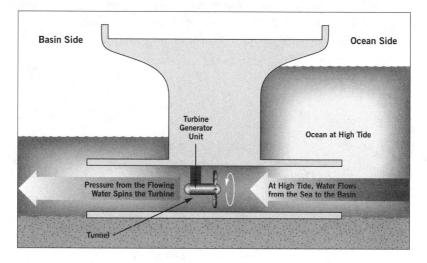

Examples of Tidal Power Plants 〰️

- **East River, New York**—Harnesses enough power for 10,000 homes
- **Land Installed Marine Powered Energy Transformer (LIMPET)**—An experimental prototype that operates by the following process:
    1. Waves push into a chamber of air.
    2. Compressed air is forced through a small hole at the turbine, turning the turbine as it is released.

Hydrogen Cells ⌒

Hydrogen cells generate electricity by splitting hydrogen into protons and electrons that are then drawn to the other side of the cell through two pathways. The protons simply travel through an electrolyte, but the electrons travel through a circuit to create electrical current. Once they reach the other side, they reunite with the protons (and oxygen) to form water.

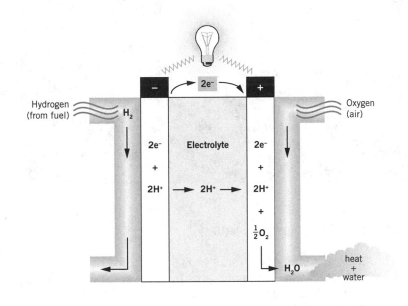

Did You Know?

Hydrogen is obtained from fossil fuels by a process called reforming.

Energy Conservation ❗

The goal is to reduce our use of fossil fuels as well as reduce the impact we have on the environment as we produce and use energy. In addition to having many negative effects on the environment, fossil fuels are a nonrenewable energy source, which means they will eventually run out. In the long term, renewable energy sources are the best bet.

Reduction in the use of both renewable and nonrenewable energy supplies can be done with greater energy conservation and improved technology efficiencies.

Examples of Energy Conservation ❗

Electric Cars	Propane or Natural Gas Cars	Cooking Oils	Mass Transit

Electric Cars	Propane or Natural Gas Cars	Cooking Oils	Mass Transit
Gaining in popularity and acceptance Good gas mileage Produce far less CO_2 pollution Hybrid vehicles have 2 motors: one electric and one gasoline-powered. Electric motor powers the car from 0–35 mph Gasoline engine takes over above 35 mph Energy is transferred from the brakes to recharge the electric motor's battery.	• Not as common as hybrids • Generate only CO_2 and water as emissions • Good gas mileage • Not many refueling stations are available.	• Can be used as alternative fuels • Oils in deep-fat fryers can be filtered and burned in diesel-fueled cars, trucks, and buses. • Start and stop the engine on pure diesel fuel • Switch to biofuel to drive	• Buses, trains, streetcars, and subways can move many more people than cars. • Generates far less pollution per person than a car

Corporate Average Fuel Economy (CAFE)

- Standards adapted by the United States
- Set mile-per-gallon standards for a fleet of cars
- Goal: reduce energy consumption, fuel consumption, and emissions by increasing the fuel economy of cars and light trucks
- Expressed as mathematical functions, which were simplified in 2012
- Will result in higher purchase prices for vehicles, but a massive reduction in air pollutants (2 million tons per year!)

Chapter 5 Summary

- Nonrenewable energy is the primary source of energy production, using primarily fossil fuels (about 80% of the global demand) including oil, coal, and natural gas
- Nuclear energy is a non-fossil fuel, nonrenewable energy source.
- Renewable energy sources account for only 14% of global energy use. These include biofuels/biodiesel, solar, hydroelectric, wind, geothermal, and hydrogen cells.
- Reduction in the use of both renewable and nonrenewable energy supplies can be done with greater energy conservation and improved technology efficiencies.

CHAPTER 6
Pollution

This chapter discusses pollution—the effects it has on Earth and its inhabitants, the types of wastes that currently exist, and how they can be managed.

Introduction to Pollution ❗

Pollution affects human health in a variety of ways. It also has a major impact on the environment as well as global and local economics.

💬 **Resiliency** is the capacity of an ecosystem to absorb disturbance and still retain its basic structure and viability. It determines how ecological systems withstand negative effects from human activities, and whether they can still deliver ecological services for future generations.

Impacts on Human Health ❗

Toxins 💬

Toxins

Toxin
- Any substance that is inhaled, ingested, or absorbed at sufficient dosages that it damages a living organism

Toxicity
- The degree to which a toxin is biologically harmful

Effects

Acute Effect
- Caused by a short exposure to a high level of toxin
- Example: effects of a snakebite

Chronic Effect
- Results from long-term exposure to low levels of toxin

Risk

Risk
- Likelihood that a person will become ill after exposure to a toxin or pathogen

Risk Assessment
- The process of calculating and predicting risk
- The basis of many environmental, medical, and public health decisions

Risk Management
- Using strategies to reduce the amount of risk
- A major role of the U.S. Department of Public Health and Public Services

Factors of Harm 💀

Several factors affect whether a substance will be harmful or not.

Dosage amount over a period of time

Number of times of exposure

Size and/or age of the organism that is exposed

Ability of the organism to detoxify that substance

Organism's sensitivity to that substance (due to genetic predisposition or previous exposure, for example)

Synergistic effects: when more than one substance combines to cause a toxic effect that's greater than any one component

Dose-Response Analysis ❗

These curves give information on the effects or toxicity of a substance.

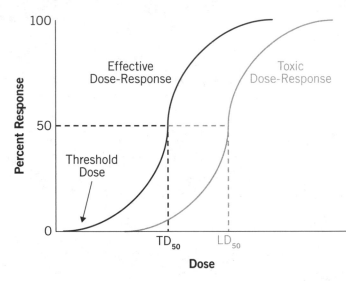

Threshold Dose	• Lowest dose at which a negative effect occurs
TD_{50}	• Toxin dose where 50% of test organisms show a negative effect from the toxin
LD_{50}	• Toxin dose that will kill 50% of the test organisms • A high LD_{50} indicates that a substance has a low toxicity. • A low LD_{50} indicates high toxicity. • A poison is any substance that has an LD_{50} of 50 mg or less per kg of body weight.

Pathogens and Human Health 🔊

Like pollution, **pathogens** can negatively affect our health. An infection is the result of a pathogen invading the body, and disease occurs when the infection causes a change in the state of health. There are several categories of pathogens.

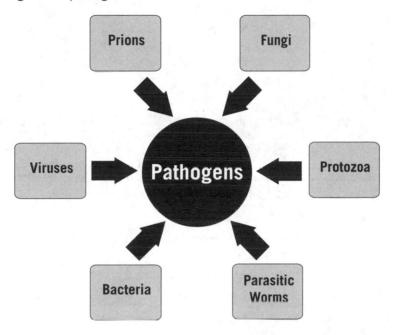

Pollution: Abbreviations and Acronyms ❗

You'll see a lot of short forms in this chapter. Here is a list of the most important ones.

BOD	Biochemical oxygen demand • A measure of the amount of oxygen that bacteria will consume while decomposing organic matter under aerobic (high O_2) conditions
BRI	Building-related illness
EPA	Environmental Protection Agency • Founded in 1970 • Established by the National Environmental Policy Act (NEPA) • An independent agency of the U.S. federal government
IPCC	Intergovernmental Panel on Climate Change
NIMBY	Not In My Backyard
OSHA	Occupational Safety and Health Association • Founded in 1971 • An agency of the U.S. Department of Labor
PAYT	Pay-As-You-Throw
SBS	Sick building syndrome
WHO	World Health Organization • Specialized agency of the United Nations • Established in 1948 • Concerned with international public health
WTE	Waste-to-Energy Program

Pollution: Key Players and Heavy Hitters 🔊

You'll see a lot of molecules and some chemistry in this chapter. Here is a list of the ones that you should be most familiar with.

CFCs	Chlorofluorocarbons
CH_4	Methane
CO	Carbon monoxide
CO_2	Carbon dioxide
DO	Dissolved oxygen
NO_2	Nitrogen dioxide
NO_x SO_x	• Several sulfur- and nitrogen-containing compounds mixed together • Both nitrogen and sulfur can combine with oxygen to make several different molecules.
O_3	Ozone
PANs	Peroxyacyl nitrates
Pb	Lead
SO_2	Sulfur dioxide
VOC	Volatile organic compound

Pollution: Legislation to Know ❗

You thought the acronyms and chemicals were bad? Well, there are even more laws and treaties for you should be familiar with for test day. Here is a summary of key legislation you should know.

Issue	Key Legislation
Environmental Issues in General	National Environmental Policy Act (NEPA)
Air Pollution	• Clean Air Act (CAA) • National Ambient Air Quality Standards (NAAQS) • State Implementation Plans (SIPs) • New Source Performance Standards (NSPS) • National Emission Standards for Hazardous Air Pollutants (NESHAPs) • Clean Air Act Amendment (CAAA) • Energy Policy Conservation Act (EPCA) • Corporate Average Fuel Economy (CAFE) • Montreal Protocol
Climate Change	• United Nations Framework Convention on Climate Change (UNFCCC) • Kyoto Protocol
Water Pollution	• Clean Water Act (CWA) • Ocean Dumping Act • Safe Drinking Water Act • Oil Spill Prevention and Liability Act
Solid and Hazardous Wastes	• Hazardous Materials Transportation Act • Resource Conservation and Recovery Act (RCRA) • Toxic Substances Control Act (TOSCA) • Comprehensive Environmental Response, Compensation, and Liability Act (CERCLA) or Superfund Program • Nuclear Waste Policy Act
Noise Pollution	• Noise Control Act • Quiet Communities Act

Air Pollution 🔔

Air pollution occurs when harmful or excessive quantities of substances including gases, particulates, and biological molecules are introduced into Earth's atmosphere. Humans have been adding to air pollution throughout our history.

Air Pollutants 🔔

There are several ways to classify air pollutants: where they come from, when they are formed, and their source location. These classifications will be explained in greater detail in the coming pages. The effects of air pollution on humans can range in severity from simply aggravating to lethal.

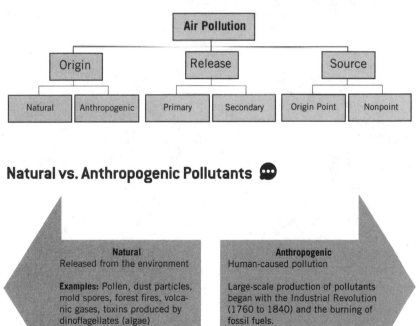

Natural vs. Anthropogenic Pollutants 💬

Natural
Released from the environment

Examples: Pollen, dust particles, mold spores, forest fires, volcanic gases, toxins produced by dinoflagellates (algae)

Anthropogenic
Human-caused pollution

Large-scale production of pollutants began with the Industrial Revolution (1760 to 1840) and the burning of fossil fuels.

Primary vs. Secondary Pollutants 💬

Primary (1°) Pollutants
Released directly into the lower atmosphere (troposphere) and are toxic

Examples: Carbon monoxide (CO)

Secondary (2°) Pollutants
Formed by the combination of 1° pollutants in the atmosphere

Examples: Acid rain is produced when sulfur oxides (such as SO_2 and SO_3) combine with water vapor

Point Source vs. Nonpoint Source Pollutants 💬

Point Source Pollution
From a specific location or stationary source

Examples: Factories; power plants; sites where wood is being burned

Nonpoint Source Pollution
Pollution that does not have a specific point of release. It comes from a combination of many mobile or moving sources.

Examples: Cows releasing methane gas within a few square miles; cars burning fossil fuels

Ask Yourself...

The world's volcanoes generate approximately 200 million tons of CO_2 each year. Is this an example of a natural or anthropogenic pollutant? Primary or secondary pollutant? Point source or nonpoint source pollutant?

EPA's Six Criteria Pollutants ❗

This dirty half dozen does the most harm to human health and welfare.

Criteria Pollutant	Key Points ❗	Cool Facts 🗨
Carbon monoxide (CO)	• Odorless, colorless gas • Released as a by-product of incompletely burned organic material, such as fossil fuels • Hazardous to human health: binds irreversibly to hemoglobin in the blood	• Time for some biology! Hemoglobin is found in red blood cells. It transports and delivers O_2 to cells. • Hemoglobin binds to CO more easily than it binds to O_2, which means oxygen isn't delivered to cells.
Lead (Pb)	• Released into the air as a particulate, during industrial smelting • Settles on land and water • Incorporated into the food chain (biomagnification) • Can cause nervous system disorders, including intellectual disabilities in children	• "Lead" in your pencil is mineral graphite, not the element lead. • It received the name "lead" because of its lead-like color when transferred to paper.
Ozone (O_3)	• Only refers to ozone that's formed as a result of human activity (tropospheric ozone) • 2° pollutant: Formed via the interaction of nitrogen oxides, heat, sunlight, and volatile organic compounds • Precursor to other 2° air pollutants • Major component of smog • Powerful respiratory irritant	• Stratospheric ozone is not a pollutant. • It absorbs UV light from the Sun and therefore protects life on our planet.

Nitrogen dioxide (NO_2)	• 2° pollutant: Formed when atmospheric nitrogen and oxygen react at high temperatures (e.g., combustion reactions in engines or industrial combustion) • Component of smog • Component of acid precipitation	More than half of the nitrogen oxides in the atmosphere are released as a result of combustion engines.
Sulfur dioxide, (SO_2)	• Colorless gas • Penetrating and suffocating odor • Released into air via coal combustion, metal smelting, paper pulping, burning of fossil fuels • Powerful respiratory irritant • Reacts with water vapor to form acid precipitation	Can also be a component of indoor pollution as a result of gas heaters, improperly vented gas ranges, and tobacco smoke
Particulates	• Exists in the form of small particles of solid or liquid material • Light enough to be carried on air currents • Respiratory irritants • Examples: soot, sulfate aerosols	Like lead, these are not gases.
Volatile organic compounds (VOCs)	• Released as a result of various industrial processes (dry cleaning, industrial solvents, using propane) • Examples: benzene, terpenes, formaldehyde • React in the atmosphere with other gases to form O_3 • Major contributor to smog in urban areas	These are not yet on the EPA's list of dirty half dozen but are a growing concern.

Did You Know?

Remember!

CO, O_3, NO_2, and SO_2 are gases. Pb and particulates are not.

Smog ❗

Smog is a kind of air pollution, originally named for the mixture of smoke and fog in the air. It can cause damage to our respiratory and cardiovascular systems.

Industrial (Gray) Smog ❗

Industrial smog, also known as gray smog, is formed from pollutants that are released when oil or coal is burned.

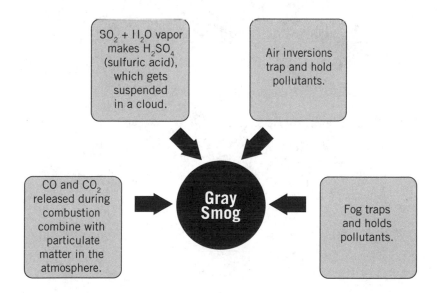

SO₂ + H₂O vapor makes H₂SO₄ (sulfuric acid), which gets suspended in a cloud.

Air inversions trap and hold pollutants.

CO and CO₂ released during combustion combine with particulate matter in the atmosphere.

Gray Smog

Fog traps and holds pollutants.

Gray smog can be deadly. In London in 1911, more than 2,000 people died in a prolonged smog incident. In 1952, about 10,000 London residents died from pneumonia, tuberculosis, heart failure, and bronchitis. This disaster prompted the Clean Air Act of 1956 in England.

Photochemical (Brown) Smog ❗

Photochemical smog, or brown smog, is formed on hot, sunny days in urban areas.

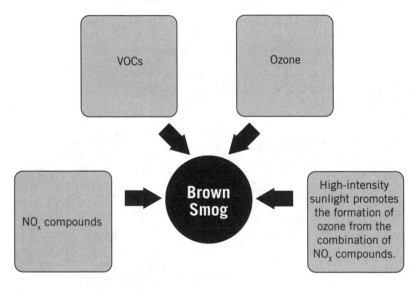

Photochemical Smog Reactions 💬

1. $2NO + O_2 \rightarrow 2NO_2$
2. $NO_2 + UV \text{ light} \rightarrow NO + O$
3. $O + O_2 \rightarrow O_3$
4. $\text{Hydrocarbons} + O_2 + NO_2 \rightarrow PANs$

How Do These Reactions Contribute to Smog? Why Are They Dangerous? 💬

- NO_2 causes a brownish haze.
- O_3 in the troposphere is very hazardous to plants, animals, and materials.
- PANs are peroxyacyl nitrates, and they cause burning eyes and damage vegetation.

Los Angeles, California, and Athens, Greece, are two cities that are particularly susceptible to photochemical smog.

Air Pollution Legislation ❗

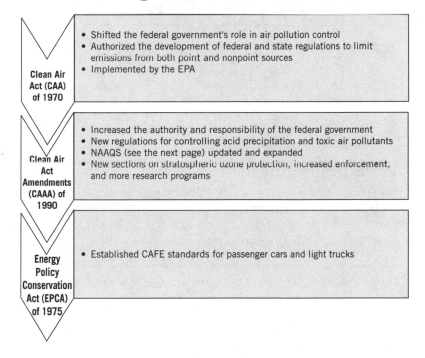

Clean Air Act (CAA) of 1970
- Shifted the federal government's role in air pollution control
- Authorized the development of federal and state regulations to limit emissions from both point and nonpoint sources
- Implemented by the EPA

Clean Air Act Amendments (CAAA) of 1990
- Increased the authority and responsibility of the federal government
- New regulations for controlling acid precipitation and toxic air pollutants
- NAAQS (see the next page) updated and expanded
- New sections on stratospheric ozone protection, increased enforcement, and more research programs

Energy Policy Conservation Act (EPCA) of 1975
- Established CAFE standards for passenger cars and light trucks

 Remember!

We reviewed CAFE standards in Chapter 5; go back there now if you need a refresher!

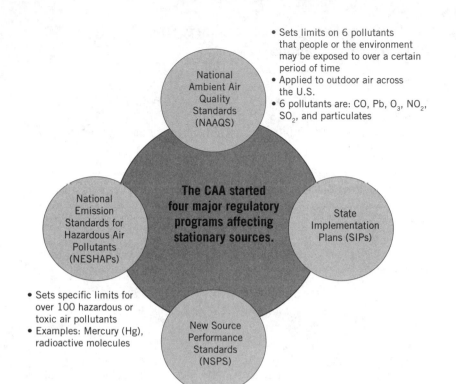

National Ambient Air Quality Standards (NAAQS)

- Sets limits on 6 pollutants that people or the environment may be exposed to over a certain period of time
- Applied to outdoor air across the U.S.
- 6 pollutants are: CO, Pb, O_3, NO_2, SO_2, and particulates

The CAA started four major regulatory programs affecting stationary sources.

National Emission Standards for Hazardous Air Pollutants (NESHAPs)

State Implementation Plans (SIPs)

- Sets specific limits for over 100 hazardous or toxic air pollutants
- Examples: Mercury (Hg), radioactive molecules

New Source Performance Standards (NSPS)

Ozone Depletion ❗

Stratospheric ozone blocks about 95% of the Sun's ultraviolet radiation (UV) and protects surface-dwelling organisms from UV damage. Declining stratospheric ozone levels were first observed in the 1950s and were due to **CFCs**, a group of manmade chemicals.

Ozone is naturally created by the interaction of sunlight and atmospheric oxygen:

$$O_2 + UV \text{ (sunlight)} \rightarrow O + O$$
$$O + O_2 \rightarrow O_3$$

Chlorofluorocarbons (CFCs)

- Invented in the 1930s
- Used in propellants, fire extinguishers, and hairspray (cans)
- Migrate to the stratosphere through atmospheric mixing
- Broken apart by intense UV radiation in the upper stratosphere
- Very stable
- Releases chlorine atoms
- $Cl + O_3 \rightarrow ClO + O_2$

 *Did You Know?*

The Antarctic continent is exposed to the greatest amount of UV radiation, but prevailing winds carry ozone-depleted air to South America, Australia, and southern Africa.

Effects of Ozone Loss

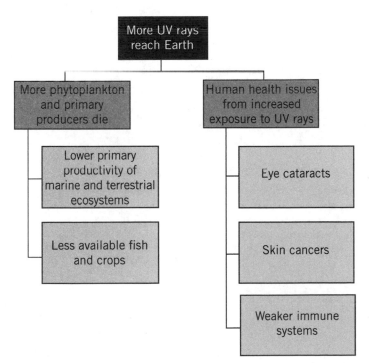

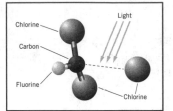

- The United States stopped production of CFCs in 1995.
- Release of ozone-depleting chemicals has been reduced by 95%.

- However, many nations still rely on CFCs.
- Research is underway to develop safe and effective substitutes.

In 1987, 146 nations signed the Montreal Protocol. This protocol called for the worldwide end of CFC production.

Acid Precipitation ❗

Acid rain, acid hail, and acid snow are all types of acid precipitation. They are caused by pollution in the atmosphere, primarily SO_2 and nitrogen oxides.

SO_2 and nitrogen oxide gases combine with water to form acids (such as nitric acid and sulfuric acid).

Deposition is usually delayed for 4 to 14 days after emission.

Atmospheric acids can travel in air currents to locations that are many miles downwind of the emission source.

Acid falls to the Earth as precipitation.

Rain usually has a pH of about 5.6 (because of the carbonic acid in rainwater).

Acid rain can have a pH as low as 2.3.

The pH scale is logarithmic; each whole pH value below 7 is ten times more acidic than the previous higher value.

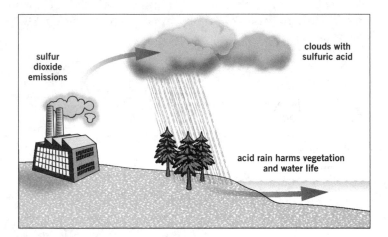

Effects of Acid Precipitation ❗

- Leaching minerals from soil (which alters soil chemistry)
- Creating a buildup of sulfur and nitrogen ions in soil

Increasing the aluminum concentration in soil to levels that are toxic for plants

Leaching calcium ions from the needles of conifers

- Elevating the aluminum concentration in lakes to levels that are toxic to fish
- Lowering the pH of streams, rivers, ponds, and lakes, which may lead to death of fish

Causing human respiratory irritation

Damaging all types of rocks, including statues, monuments, and buildings

Which Areas Are Particularly Vulnerable to Acid Precipitation? ❗

- Areas with already acidic soils
- Areas with soil derived from granite
- Areas where the soil has been leached of its natural calcium content

---

 Did You Know?

- Calcium acts as a natural buffer and tempers the effects of acid precipitation.
- Acid rain has dissolved much of the limestone rock that Florida sits on; this is a significant cause of sinkholes in Florida.

What's Being Done? ❗

In some areas of the world, progress has been made toward controlling acid precipitation.

CAA, CAAA, and NAAQS have helped decrease air pollution, especially SO_2 and NO_x emitted from industrial plants.

Less acid rain

Some areas continue to be damaged by acid rain.

Many ecosystems will not be able to continue to tolerate significant lowering of their pH.

Motor Vehicles and Air Pollution 🛑

Passenger motor vehicles are a major contributor to pollution. They produce significant amounts of nitrogen oxides, carbon monoxide, and hydrocarbons.

Catalytic Converter

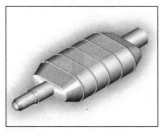

- Platinum-coated device
- Controls exhaust emissions
- Used in internal combustion engines (fueled by either gasoline or diesel), kerosene heaters, and stoves
- Increasing use since the mid-1970s
- Oxidize most of VOCs
- Oxidize some CO
- Converts them to CO_2
- Without a catalytic converter, these pollutants would be emitted in exhaust.
- Newer models of catalytic converters also reduce some nitrogen oxides.

Vehicles of the Future 🛑

More environmentally friendly vehicles are being developed, and people are investing in them.

Electric Vehicles	Hybrid Vehicles	Hydrogen Fuel Cell Vehicle
• Plug and drive! Powered by an electric motor and battery • Do not burn gasoline or diesel • Zero tailpipe emissions • First of these had a limited traveling range and lacked amenities like air conditioning. • New battery technology has reduced their weight and cost. • Most can now travel 200–300 km on a full charge. • Have a promising future! • Examples: Nissan Leaf, Ford Focus, Kia Soul, Chevrolet Bolt, Volkswagen e-Golf, Hyundai IONIQ, BMW i3 and Tesla Model X	• Use both electricity and gasoline • Can travel 20–80 km (depending on model) on a full charge • Most day-to-day driving is done on the electric battery. • Once the battery is used up, a gasoline engine or generator turns on for an additional 500+ km of gasoline range. • Examples: Toyota Prius, Ford Fusion, Chevrolet Volt, Honda Clarity, Audi A3, Kia Optima, Hyundai Sonata, Chrysler Pacifica and BMW 330	• Would produce even less pollution than a hybrid vehicle • However, hydrogen fuel cells are very expensive. • These cars aren't yet economically viable. • If you need a review of hydrogen fuel cells, go back to Chapter 5.

Government regulations, incentives (such as tax credits), and afford-ability will determine how large of a sector these vehicles will claim of the vehicle market.

Indoor Air Pollution ❗

The concept of indoor air pollution is relatively new, but it's an impor-tant one. According to the WHO, indoor air pollution is responsible for 4 million annual deaths worldwide. The EPA lists indoor air pollution as one of five major environmental risks to human health.

Air pollutants are usually at a higher concentration indoors than outside.

What Causes Air Pollution?

- Pollutants from the outside move inside when doors and windows are opened.
- Pollutants remain trapped inside until air currents move them out a door, window, or ventilation system.

Developed Countries

- Humans spend a significant amount of time indoors.
- We work and live in well-sealed buildings with little air exchange.

Developing Countries

- Dung, wood, and crop waste are used as fuel to heat homes and to cook food.
- Indoor burning leads to dangerous levels of particulate matter.

Indoor Air Pollutants ❗

Volatile Organic Compounds (VOCs)
- Found in carpet, furniture, plastic, oils, paints, adhesives, pesticides, and cleaning fluids

Tobacco Smoke
- Leading cause of lung cancer
- Affects the health of the smoker and those around them

Second-hand Smoke
- Causes the same symptoms in non-smokers who breathe in second-hand smoke
- Contains over 4,000 different chemicals
- Classified as a Group A carcinogen (meaning that it causes cancer in humans) by the EPA

Carbon Monoxide (CO)
- Comes from gas leaks or poor gas combustion devices
- CO detectors are available for homes and can prevent CO poisoning.

Radon Gas
- Second leading cause of lung cancer (after smoking) in the United States
- Emitted by uranium as it undergoes radioactive decay
- Seeps up through rocks and soil, and then enters buildings
- Not found everywhere; must be tested for specifically
- Homes that were built after 1990 have radon-resistant features.

Biotic Pollutants
- Tiny insects, fungi, and bacteria
- Cause allergies and/or asthma in many people
- Example: Bacteria growing in an air conditioner water tank are distributed throughout a house with cool air.

Sick Building Syndrome (SBS)
- This occurs when the majority of a building's occupants experience certain symptoms.
- Results vary with the amount of time spent in the building.
- SBS is difficult to diagnose, and the specific cause is difficult to identify.
- WHO reported in 1984 that up to 30% of buildings worldwide had poor indoor air quality.

Building-Related Illness (BRI)
- This occurs when signs and symptoms can be attributed to a specific infectious organism that resides in a building.
- Common symptoms include
  ○ irritation of the eyes, nose, and throat
  ○ a change in odor or taste sensitivity
  ○ neurological symptoms: headaches, dizziness, reduction in the ability to concentrate, memory loss
  ○ nausea or vomiting
  ○ skin irritation

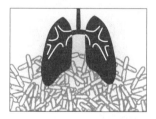

Example: Legionnaires' Disease
- Caused by inhaling *Legionella pneumophila* bacteria
- Severe form of pneumonia or lung inflammation
- Non-communicable: cannot be passed from person to person

How to Avoid Indoor Air Pollution 🔊

- Quit smoking
- Limit exposure to pesticides or cleaning fluids
- Ensure buildings are well ventilated
- Test for and/or block radon

Thermal Pollution ❗

You should know two types of thermal pollution: the **urban heat island effect** and **thermal inversions.**

Urban Heat Island Effect ❗

Urban environments are generally warmer than the countryside that surrounds them. This heat difference can be as small as a few degrees or up to a difference of 22°F (12°C)!

Temperature Profile of an Urban Heat Island

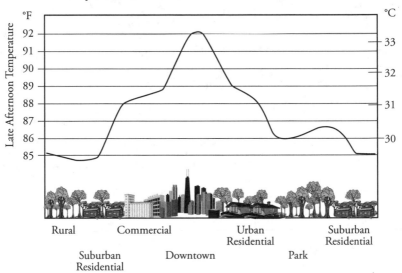

What Causes Urban Heat Islands? ❗

- Buildings, concrete, and asphalt absorb and then radiate heat.
- Industrial and domestic machines directly warm the air.
- Minimal green spaces means minimal cooling effects.

The high temperatures of heat islands increase the rates of photochemical reactions, which in turn leads to photochemical smog.

How Can Urban Heat Islands Be Reduced?

1. Replace dark, heat-absorbing surfaces (such as roofs) with light-colored, heat-reflecting surfaces
2. Plant trees and add to green spaces

How Do Green Spaces Cool Urban Heat Islands?

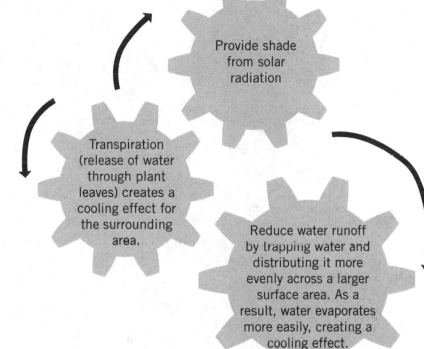

Provide shade from solar radiation

Transpiration (release of water through plant leaves) creates a cooling effect for the surrounding area.

Reduce water runoff by trapping water and distributing it more evenly across a larger surface area. As a result, water evaporates more easily, creating a cooling effect.

Did You Know?

Evaporating water has a cooling effect. Concrete and asphalt in cities increase water runoff. This creates deep pools of water and less water evaporation.

Pollution

Green Roofs 💬

Green roofs on city buildings can help decrease thermal pollution. A **green roof,** or living roof, is fully or partially covered with plants, greenery, gardens, and other vegetation planted over some type of waterproofing material.

Benefits of a Green Roof 💬

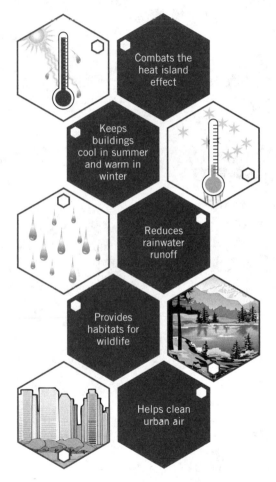

Thermal Inversion 🛑

Temperature inversions, or **thermal inversions,** can occur over any city where large masses of warm air can become stalled. They are another type of thermal pollution.

In normal atmospheric conditions, warm and polluted air over a city rises into the cooler atmosphere. This happens because warm air is less dense than the surrounding cool air, and less dense objects float.

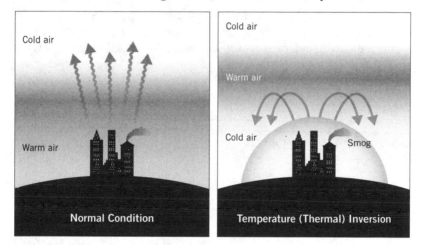

What Causes Thermal Inversions? 🛑

Thermal inversions are caused by warm air above the city, which blocks polluted air from rising into the atmosphere. As such, polluted air remains hanging above the city, which can lead to respiratory and other health issues.

Inversions often occur in cities surrounded by mountains or cities bordered by mountains on one side and ocean on the other. Los Angeles and Beirut are two examples.

Water Pollution ❗

Agricultural, industrial, and mining activities all contribute to water pollution. Similar to sources of air pollution, sources of water pollution are classified as point or nonpoint.

 Remember!

We reviewed point and nonpoint sources of pollution earlier in this chapter. Point source pollution comes from a specific location or stationary source. In contrast, nonpoint source pollution does not have a specific point of release. It comes from a combination of many mobile or moving sources.

Major Sources of Water Pollution ⚠️

Excess nutrients (such as nitrogen or phosphate)

Organic or human waste

Toxic waste (such as pesticides, petroleum products, heavy metals, acids)

Sediment, or soil washed with runoff water into streams

Hot water discharged from industrial facilities, where it was used as a coolant

Cold water from dam releases, discharging it from the bottom of a reservoir

Coliform bacteria, found in the intestines of animals

Invasive species (such as zebra mussels)

Thermal pollution

The presence of Coliform bacteria in water indicates the presence of fecal matter.

Infectious agents can be found in human and animal waste. Fecal waste can contain disease-causing bacteria (as well as symbiotic bacteria that aid in digestion). Several human diseases (such as cholera and typhoid fever) are caused by human waste entering a water source.

 ## Did You Know?

Approximately one-third of male fish in British rivers are in the process of changing sex due to pollution. Hormones in human sewage, including those produced by the female contraceptive pill, are thought to be the main cause.

Water Pollution Outcomes 🛑

Different bodies of water recover from pollution in different ways. Water pollution can result in large dead zones (see the next page).

Recovering from Water Pollution 🛑

Standing Bodies of Water (Ponds and Lakes)	Groundwater	Flowing Streams and Rivers	Oceans
• Do not recover quickly • Minimal water flow • Pollutants not diluted • Pollutants undergo biomagnification	• Does not recover quickly • Minimal water flow • Minimal flushing, mixing, and dilution • Very cold and low in dissolved oxygen, so slow to recover from degradable waste	• Movement of water allows flushing, mixing, and dilution • Can recover from moderate levels of degradable pollutants	• Can dilute, flush, and decompose large amounts of degradable waste • Long-term capacity for recovery is unknown

What Is a Dead Zone and How Does It Form? ❗

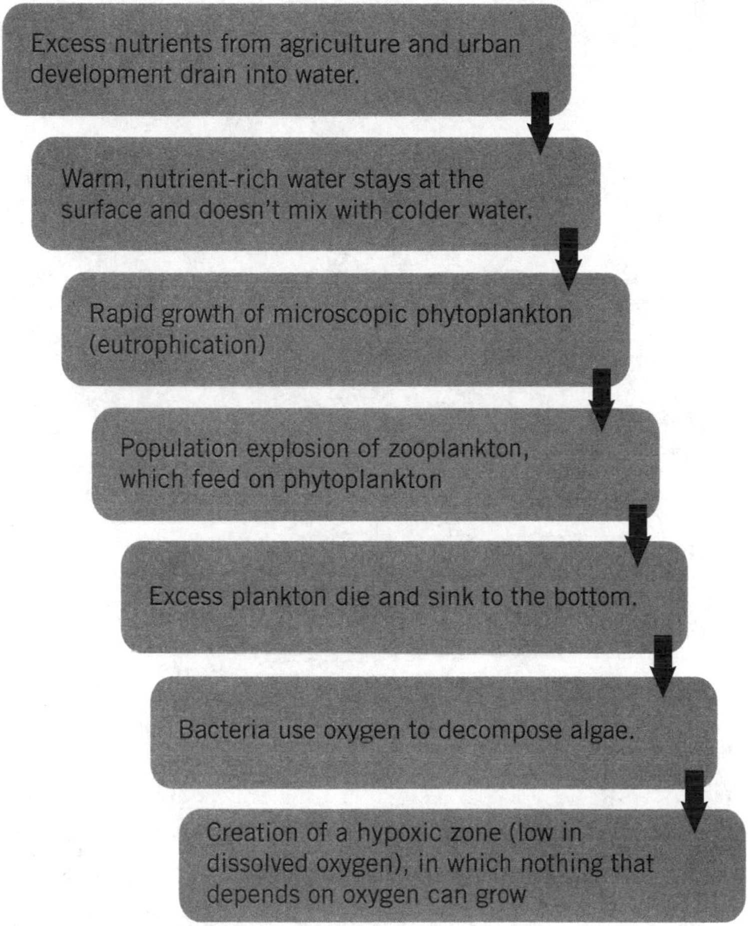

Excess nutrients from agriculture and urban development drain into water.

⬇

Warm, nutrient-rich water stays at the surface and doesn't mix with colder water.

⬇

Rapid growth of microscopic phytoplankton (eutrophication)

⬇

Population explosion of zooplankton, which feed on phytoplankton

⬇

Excess plankton die and sink to the bottom.

⬇

Bacteria use oxygen to decompose algae.

⬇

Creation of a hypoxic zone (low in dissolved oxygen), in which nothing that depends on oxygen can grow

---

Dead zones can be broken up in colder, wetter weather.

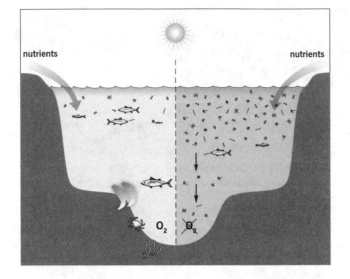

Mississippi Dead Zone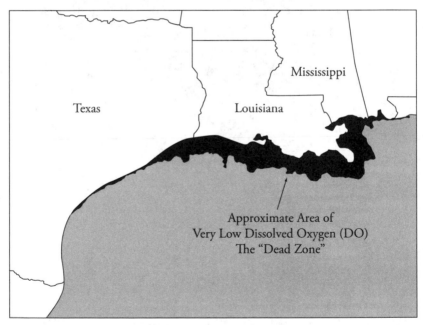

Approximate Area of
Very Low Dissolved Oxygen (DO)
The "Dead Zone"

This dead zone has caused the collapse of the shrimp and shellfish industries in that region.

How Do We Deal with Water Pollution? ⚠

1. Reducing or removing the sources of pollution

2. Treat the water in order to remove pollutants or render them harmless in some way

Water Quality ⚠

Water quality tests can be performed on water samples. They test for the presence of various chemicals and indicator species.

Important Factors in Judging Water Quality ❗

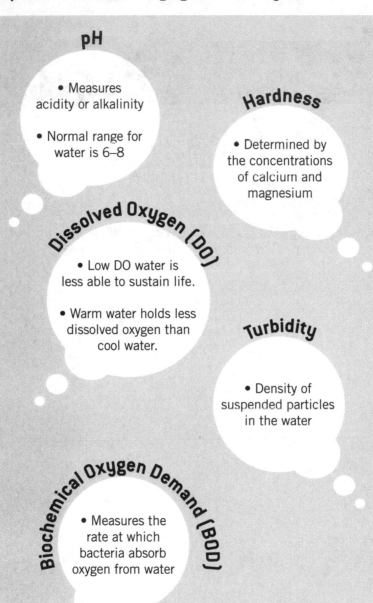

pH

- Measures acidity or alkalinity

- Normal range for water is 6–8

Hardness

- Determined by the concentrations of calcium and magnesium

Dissolved Oxygen (DO)

- Low DO water is less able to sustain life.

- Warm water holds less dissolved oxygen than cool water.

Turbidity

- Density of suspended particles in the water

Biochemical Oxygen Demand (BOD)

- Measures the rate at which bacteria absorb oxygen from water

Aquatic Invertebrates	Fish and Water Oxygen	Biodiversity

- Some species of aquatic invertebrates can survive in polluted water, e.g., rat-tailed maggots and sludge worms.
- Others have very low tolerance for any pollution (sensitive); e.g., freshwater shrimp and mayfly larva.
- Lack of sensitive species suggests water is polluted.

- Fish rely heavily on dissolved oxygen in water to survive.
- If oxygen levels are low due to pollution, few varieties of fish will survive.
- Scientists monitor how fish populations change over time.

- The collective number of species can be used as an indicator of environmental health.
- Polluted environments do not support a diverse range of species.

Wastewater 🔊

Once water has been used by humans, it is called **wastewater**. This includes storm water runoff, human sewage (water drained from showers, sinks, toilets, dishwashers, and washing machines), and water from industrial processes.

Storm water goes into storm drains and is then dumped directly into rivers, without any treatment. Other wastewater is transported to municipal sewage treatment plants via sewage pipes.

How Do Municipal Sewage Treatment Plants Work? ❗

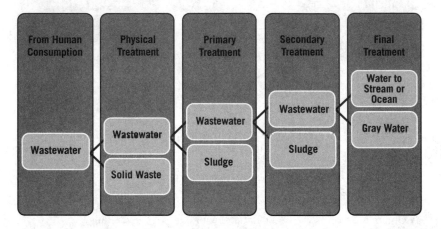

Process	What Happens?	What Is Removed?
Physical Treatment	• Water is filtered through screens.	• Solid waste or debris • Examples: stones, sticks, rags, toys
Primary Treatment	• Water is sent to a settling tank. • Suspended solids settle out as sludge. • Chemically treated polymers can be added to help this happen.	• 60% of suspended solids • 30% of organic waste
Secondary Treatment	• Biological treatment to continue removing biodegradable waste • Trickling filters: aerobic bacteria digest waste as it seeps over bacteria-covered rock beds • Activated sludge processor: water is pumped into a tank filled with aerobic bacteria	• 37% of suspended solids • 65% of organic waste • 70% of toxic metals • 70% of organic chemicals • 70% of phosphates • 50% of nitrogen • 5% of dissolved salts
Final Treatment	• Chlorination (inexpensive but can form potential carcinogens) • Ozonation and UV radiation (more expensive and safer, but less effective)	• Any remaining living cells

With primary and secondary treatment, 97% of suspended solids and 95–97% of organic waste is removed from wastewater. Note that wastewater treatments typically do not remove radioactive isotopes or persistent organic chemicals (such as pesticides).

What Ends Up Where?

- Solid waste or debris is separated and sent to a landfill.
- Sludge is further processed with anaerobic bacteria, and then used as fertilizer.
- Processed water is discharged into a stream, the ocean, or used as gray water.
- Gray water is used to water lawns, golf courses, and plants in nurseries.

Did You Know?

In many developing countries, sewage water is dumped into the nearest river or ocean. Dealing with large amounts of waste in this way poses serious risks to human health and the health of the aquatic ecosystems.

Tertiary Treatment

Water is passed through a series of sand and carbon filters.

Further chlorination

Adds more expense after primary and secondary treatment

Helps arid or semi-arid regions reclaim as much water as possible

Private wastewater treatment

Environmentally friendly type of waste disposal

Primary and secondary treatments similar to municipal treatment plants

Water discharged into leachate (drain) fields, where soil percolates the water

Water Quality Legislation ❗

The 1972 Clean Water Act (CWA) was created in response to problems arising from polluted bodies of water in the late 1960s. It has had dramatic positive effects on the quality of water in the United States. Several other important pieces of federal law also cover and protect water quality.

Date	Name of Legislation	What It Did
1972	Clean Water Act	• Used regulatory and non-regulatory tools to protect all surface waters in the United States • Sharply reduced direct pollutant discharge into waterways • Financed municipal wastewater treatment facilities to manage polluted runoff • Achieved the broader goal of restoring and maintaining the chemical, physical, and biological integrity of the nation's waters • Supported "the protection and propagation of fish, shellfish, and wildlife, and recreation in and on the water"
1972	Ocean Dumping Act	Made it unlawful for any person to dump, or transport for the purpose of dumping, sewage sludge, or industrial waste into ocean waters
1974, 1996, 2005, 2011, 2015	Safe Drinking Water Act	Established a federal program to monitor and increase the safety of the drinking water supply. It does not apply to wells that supply fewer than 25 people. Amendments in recent years have led to more stringent regulation of lead and algal toxins in drinking water.
1990	Oil Spill Prevention and Liability Act	Strengthened the EPA's ability to prevent and respond to catastrophic oil spills. Established a trust fund (financed by a tax on oil), which is available to clean up spills.

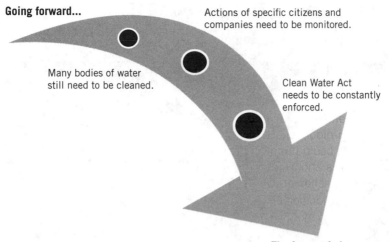

Going forward...

Actions of specific citizens and companies need to be monitored.

Many bodies of water still need to be cleaned.

Clean Water Act needs to be constantly enforced.

The future of clean water!

Did You Know?

Experts say that Americans have some of the cleanest drinking (tap) water in the world.

Solid Waste

Garbage is a more common word for solid waste, which can come from either municipal or industrial sources. Many types of garbage are hazardous in that they are a threat to human health and the environment.

Reduce, Reuse, Recycle, and Compost ❗

These four processes help minimize solid waste:

Reduce	Reuse	Recycle	Compost

• Minimize disposable waste
• Many types of packaging are extremely (and often needlessly) wasteful
	• Applies to products that are fundamentally disposable, but can be used over and over again
	• Prevents high-quality goods from becoming waste
	• Examples:
	  • Refillable bottles
	  • Reusable packing materials
	  • Secondhand goods
	  • Cloth shopping bags
		• Reuse of materials
		• Two types: primary recycling and secondary recycling (see below)
			• Allows organic material in solid waste to be decomposed and reintroduced into the soil

Types of Recycling 💬

Primary Recycling
• Materials such as plastic and aluminum are used to rebuild the same product.
• **Example:** Aluminum from recycled cans is used to produce more aluminum cans.

Secondary Recycling
• Materials are reused to form new products that are usually lower quality goods.
• **Examples:** Old tires are recycled to make carpet; plastic bottles are recycled to create decking material.

How Much Are We Recycling? 💬

Here is how much of each of the following materials was recycled in the year 2014:

Material	Percent of This Material That Was Recycled in 2014
Lead-acid batteries	98.9
Steel cans	70.7
Newspaper	67.0
Yard trimmings	61.1
Aluminum beer and soda cans	55.1
Tires	40.5
Electronics	41.7
Glass	32.5
PET bottles and jars	31.2

PAYT Programs 🔊

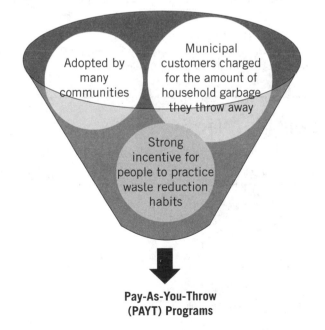

Pay-As-You-Throw
(PAYT) Programs

Landfills !

Modern landfills are very different from the traditional caricature of a garbage dump filled with heaps of junked cars and rats foraging for food scraps. Federal regulations that protect human health and the environment have paved the way for sanitary landfills.

Landfill Guidelines and Regulations !

✓ Landfills cannot be located near geological faults, wetlands, or flood plains.
✓ Landfills are lined with geomembranes, plastic sheets that form a protective barrier.
✓ The perimeter of landfills is lined with two feet of clay.
✓ Waste is frequently covered with soil to control insects, bacteria, rodents, and odor.
✓ Decomposed material that percolates to the bottom of the landfill (called leachate) is piped to the top of the site, collected, and closely monitored.
✓ Gases from the landfill (like methane) can be piped up and used to generate electricity.
✓ Landfills must be positioned at least six feet above the water table.
✓ Groundwater is tested frequently for quality.
✓ When a hole is full, it is capped with an engineered cover, monitored, and provided with long-term care.

 Did You Know?

About one-third of an average dump is made up of packaging material.

Structure of a Modern Landfill

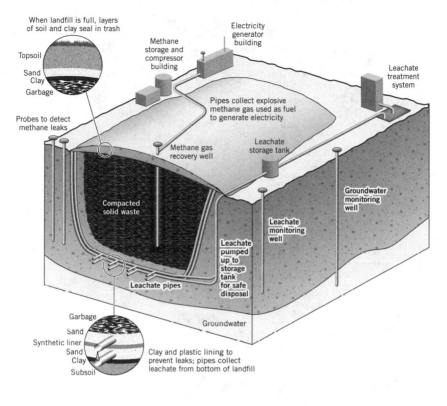

When landfill is full, layers of soil and clay seal in trash

Topsoil
Sand
Clay
Garbage

Probes to detect methane leaks

Methane storage and compressor building

Electricity generator building

Leachate treatment system

Pipes collect explosive methane gas used as fuel to generate electricity

Methane gas recovery well

Leachate storage tank

Compacted solid waste

Groundwater monitoring well

Leachate monitoring well

Leachate pumped up to storage tank for safe disposal

Leachate pipes

Garbage
Sand
Synthetic liner
Sand
Clay
Subsoil

Groundwater

Clay and plastic lining to prevent leaks; pipes collect leachate from bottom of landfill

---

The acronym NIMBY (Not in My Backyard) has been around since the mid-20th century, but it gained popularity in the 1980s when the U.S. experienced a widely publicized garbage crisis. While people agreed that landfills were needed, no one wanted a landfill close to their home.

Energy released from the incineration can be used to generate electricity.

Waste is burned in municipal incinerators (after recyclables are removed).

WTE programs are effective in large municipal areas, where waste needs to be transported only short distances.

**Waste-to-Energy
(WTE) Programs**

Hazardous Waste 🛈

Hazardous waste is any waste that poses a danger to human health. It must be dealt with in a different way than other types of waste. Hazardous waste includes batteries, various household cleaners, paint, solvents, and pesticides.

Types of Hazardous Waste ❗

There are four categories of hazardous wastes:

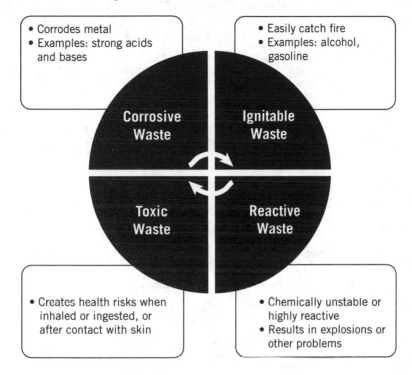

- Corrodes metal
- Examples: strong acids and bases

Corrosive Waste

- Easily catch fire
- Examples: alcohol, gasoline

Ignitable Waste

Toxic Waste

- Creates health risks when inhaled or ingested, or after contact with skin

Reactive Waste

- Chemically unstable or highly reactive
- Results in explosions or other problems

What Is Done with Hazardous Waste? ❗

Disposal Option	How It Works	Pros	Cons
Landfills	• Typically used for solid waste • Specific areas in landfills are designated for hazardous waste. • These areas have higher standards and tighter regulations.	• Easy • Regulated and controlled	• A leak can release hazardous chemicals into groundwater. Can emit harmful gases into the atmosphere
Surface Impoundments	• Typically used for liquid waste • A shallow, lined pool is created. • Hazardous liquid is added and left to evaporate.	• Easy • Regulated and controlled	• Temporary • Waste can overflow, blow out, or leak
Deep-Well Injections	• A hole is drilled into the ground, below the water table and through a layer of clay (impervious or water-resistant soil). • Waste is injected into porous rock.	• Regulated and controlled • Long-term disposal • Isolated from human contact	• Wells can become corroded and leak waste into soil or ground water.

Surface Impoundments 💬

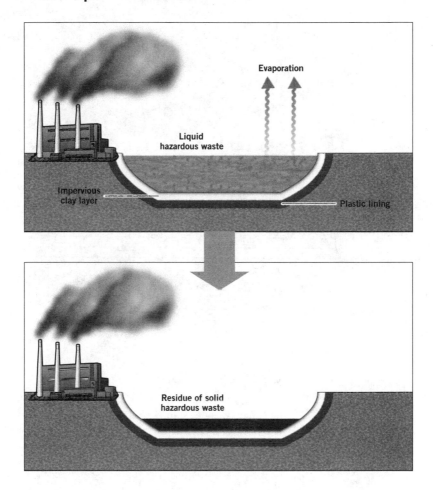

Evaporation

Liquid hazardous waste

Impervious clay layer

Plastic lining

Residue of solid hazardous waste

Deep-Well Injection 💬

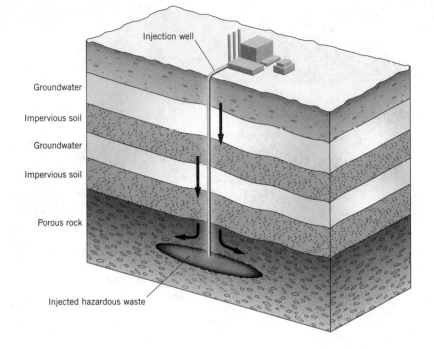

Injection well

Groundwater

Impervious soil

Groundwater

Impervious soil

Porous rock

Injected hazardous waste

Types of Radioactive Waste ❗

The EPA has defined six categories of radioactive wastes, depending on where the waste comes from.

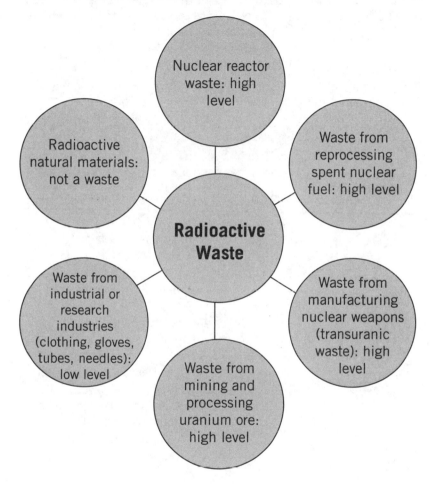

What Is Done with Radioactive Waste? ❗

Type of Radioactive Waste	Disposal Options
Low Level	• Stored on-site by licensed facilities until the radioactivity has degraded • Shipped to a low-level waste disposal facility and stored there
Mixed (contains both hazardous and radioactive waste)	• Stored on-site by licensed facilities until the radioactivity has degraded • Shipped to a low-level waste disposal facility and stored there

High Level	• Requires sophisticated and highly regulated treatment and management to successfully isolate it from the biosphere
	• Majority is currently stored at the site where it is generated
	• Some is sent to the Waste Isolation Pilot Plant (WIPP) in New Mexico
	  ○ Operational since 1999
	  ○ Managed by the U.S. Department of Energy
	• Search for and development of one major site for the disposal of all radioactive waste is still underway
	  ○ Yucca Mountain, Nevada is a possibility.
	  ○ Future site(s) will likely involve deep-geologic placement; the storage facility will be excavated deep (typically below 300 meters or 1,000 feet) within a stable geologic environment.

Contaminated Waste Sites ❗

Historical Overview ❗

After the 1970s, new regulations for the disposal of hazardous wastes improved the process for adding new wastes to landfills with minimal impact on the environment.

At the same time, a problem lingered: what should be done with sites that were already contaminated by hazardous waste or pollutants?

These sites are known as **brownfield sites,** and they had to be cleaned up.

Those who had acted irresponsibly had to be held accountable for the environmental problems they had caused.

For these reasons, the U.S. Congress created the **Superfund Program.**

Superfund Program ❗

A government program designed to clean up sites contaminated with hazardous substances and pollutants

Established in 1980

Administered by the EPA

Sites managed under this program are referred to as "Superfund sites"

Examples of Superfund Sites ❗

Rocky Flats, Colorado ❗

- U.S. government was responsible for the damage.
- From 1952 to the early 1990s, components of nuclear weapons and stainless steel were manufactured on this site.
- The area has now been significantly cleaned up and is home to a variety of plants and animals. It also acts as a wind-power testing site.

Love Canal, New York 🔔

- Land was purchased by various companies and used as a landfill in the 1940s. Thousands of tons of hazardous chemicals were dumped.
- Landfill was covered and the town purchased the land in 1953.
- About 100 homes and a school were built on the site.
- Problems began in the 1970s:
  - Rusting drums full of waste sticking up above ground
  - Dead and dying trees and plants
  - Pools of smelly liquids in home basements
  - Health issues among residents and schoolchildren
- The Environmental Protection Agency got involved.
- Hundreds of families were relocated, homes were demolished, and environmental cleanup began.
- Cleanup concluded and the site was removed from the Superfund list in 2004.

Laws for Solid and Hazardous Wastes 🔔

It was in response to the situation at Love Canal (and other similar situations) that laws like the Resource Conservation and Recovery Act (RCRA) and CERCLA* were passed. Industry produces the largest amounts of hazardous waste, and most developed countries regulate the disposal of these wastes. U.S. law mandates that hazardous materials be tracked "from cradle to grave" thanks to laws like RCRA.

CERCLA stands for Comprehensive Environmental Response, Compensation, and Liability Act, also known as Superfund.

Date	Name of Legislation	What It Did
1975	Hazardous Materials Transportation Act	Governed the transportation of hazardous material and wastes in commerce
1976	The Resource Conservation and Recovery Act	• The solid waste program encouraged states to develop comprehensive plans to manage nonhazardous industrial solid waste and municipal solid waste. It also set criteria for municipal solid waste landfills and other solid waste disposal facilities, and prohibited the open dumping of solid waste. • The hazardous waste program established a system for controlling hazardous waste from the time it is generated until its ultimate disposal—in effect, from "cradle to grave." • The underground storage tank (UST) program regulates underground storage tanks containing hazardous substances and petroleum products.
1976	Toxic Substances Control Act (TOSCA)	• This act gave the EPA the ability to track the 75,000 industrial chemicals currently produced in or imported to the United States. • EPA repeatedly screens these chemicals and can require reporting or testing of those that may pose an environmental or human health hazard. • The act also allows the EPA to ban the manufacture and import of those chemicals that pose an unreasonable risk.

1980	The Comprehensive Environmental Response, Compensation, and Liability Act (CERCLA), commonly known as Superfund	• Created a tax on the chemical and petroleum industries and provided broad federal authority to respond directly to releases or threatened releases of hazardous substances that may endanger public health or the environment • Established prohibitions and requirements concerning closed and abandoned hazardous waste sites • Provided for liability of persons responsible for releases of hazardous waste at these sites • Established a trust fund to provide for cleanup when no responsible party could be identified
1982	Nuclear Waste Policy Act	Established both the federal government's responsibility to provide a place for the permanent disposal of high-level radioactive waste and spent nuclear fuel, and the generators' responsibility to bear the costs of permanent disposal

Economic Impact ❗

Economics is the study of how people use limited resources to satisfy their wants and needs. Decisions based solely on economics are morally neutral; they do not say anything about ethics or fairness.

Term ❗	Definition ❗	Example 〰
Cost-Benefit Analysis	• Economists weigh benefits against costs • Helps make decisions about how to use resources • Can be difficult: it is hard to assign monetary value to intangible properties such as beauty or clear air	Before starting construction of the Ocklawaha River Canal across central Florida in the 1960s, a cost-benefit analysis was performed. Analysis determined that there would be a few benefits, but that the proposal would be detrimental to most interest groups (shippers, fishers, citizens, the environment). Because of this evidence, the project was halted in 1971.
Marginal Cost	Cost added by producing one additional unit of a product or service	What is the cost to the economy and/or environment of removing one more acre of forest?
Marginal Benefits	Benefit of producing one additional unit of a product or service	What is the benefit to the economy and/or cost to the environment of planting one more acre of forest?

Positive Externalities	Unwanted or unanticipated consequences that are positive	A beekeeper keeps bees for their honey. Positive externality: pollination of surrounding crops. A firm builds a large tree plantation. Positive externality: people and animals nearby benefit from this greenspace.
Negative Externalities	Unwanted or unanticipated consequences that are negative	Smokers cause second-hand smoke. Many people depend on personal vehicles to get to work, but cars produce more CO_2 per passenger kilometer than any other form of land transport.

Sustainability 😛

The capacity to survive and endure is the sustainability of a system. This is how biological systems remain diverse and productive.

Requirements of Sustainability 😛

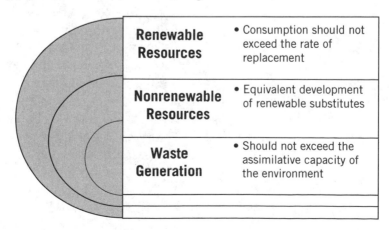

Renewable Resources	• Consumption should not exceed the rate of replacement
Nonrenewable Resources	• Equivalent development of renewable substitutes
Waste Generation	• Should not exceed the assimilative capacity of the environment

Things Are Getting Better ❗

Since the 1970s, several major changes have occurred in pollution control, which have had very positive effects.

What Has Happened	Effects
• Phasing out of lead gasoline • Introduction of car engines that burn more cleanly	Significant decreases in the atmospheric content of lead and CO
Clean Air Act and its amendment	Cars produced after 1999 emit 75% fewer pollutants than cars made before 1970.
Use of catalytic converters in engines, heaters and stoves	Decrease in CO and hydrocarbon emissions
Use of scrubbers in coal-burning plants	Reduced the amount of SO_2 released into the atmosphere
Increase in recycling and composting to deal with solid waste	Diverts millions of tons of material away from landfills and incinerators per year
Implementation and success of the Superfund program	Has contributed to cleaning up almost 1,500 sites contaminated with hazardous substances and pollutants
Clean Water Act	• Over 90% of community water systems now meet federal health standards. • Twice as many streams are sufficiently clean to allow fishing and swimming. • 80% decrease in wetland loss
Discovery and development of alternate and renewable energy sources	Decreased pollution, especially air pollution
Increase in mass transit	• Fewer vehicles on the road • Decreased air pollution

Chapter 6 Summary

- This chapter covered four main areas: air pollution, climate change, water pollution, and waste.
- Air pollution has increased dramatically since the Industrial Revolution due to use of fossil fuels. Sources of air pollution are either point source or nonpoint source.
- Major air pollutants are carbon monoxide, lead, ozone, nitrogen dioxide, sulfur dioxide, and particulates.
- Greenhouse gases are molecules that absorb heat and warm the lower atmosphere. The major greenhouse gases are carbon monoxide, nitrous oxide, and methane.
- The greatest contributor to water pollution is excess nutrients from runoff, which can lead to eutrophication.
- Factors for testing water quality are pH, hardness, dissolved oxygen, turbidity, and biological oxygen demand.
- Solid waste consists of many types that may each be a threat to human health and the environment.
- Municipal waste (produced by households) and industrial waste (from commercial production) is either recycled, composted, buried in landfills, or incinerated for Waste-to-Energy programs.
- Hazardous waste poses dangers to human health and must be dealt with specifically via deep well injections, surface impoundments, and landfills, for example.

CHAPTER 7
Global Change

As society has become more technologically advanced, human impact on the environment has increased substantially. This chapter looks at a few key issues that have resulted from global change, and which will shape environmental policy for generations to come: ozone depletion, global warming, loss of biodiversity, the need for sustainability, and globalization.

Ozone 🔔

All ozone is O_3 and is the same chemically. Ozone is naturally created by the interaction of sunlight and atmospheric oxygen.

Ozone Formation

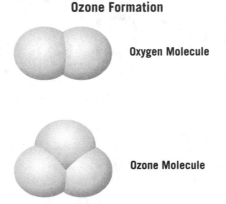

Oxygen Molecule

Ozone Molecule

The **stratospheric ozone** layer is responsible for blocking about 95% of the Sun's **ultraviolet radiation** (UV), thus protecting surface-dwelling organisms from UV damage. On the other hand, **tropospheric ozone** is a powerful respiratory irritant and precursor to secondary air pollutants.

Up high, ozone helps us; down low, it hurts us. O_3 is a secondary pollutant that is formed in the troposphere as a result of the interaction of nitrogen oxides, heat, sunlight, and volatile organic compounds (VOCs). Tropospheric ozone is a major component of what we think of as smog.

Ozone Depletion 🔔

Ozone loss is greatest in the spring as the chlorine breaks down ozone into O_2:

$$Cl + O_3 \rightarrow ClO + O_2$$

Chlorine acts as a catalyst; it is not changed by its reaction with ozone, and it can help break down another O_3 molecule immediately. As the air continues to warm, the natural production of ozone increases as more sunlight catalyzes the combination of oxygen back into ozone.

Ozone loss has negative implications for the Earth's ecosystems and human health:

- Increased number of UV rays kill phytoplankton and other primary producers
- Decreased primary productivity of both marine and terrestrial ecosystems
- Human health issues like eye cataracts, skin cancers, and weakened immune systems

Strategies for Reducing Ozone Depletion ❗

One key international agreement designed to manage ozone depletion worldwide is the Montreal Accord.

Date	Agreement	What It Did
1978	Montreal Accord	Cut the emissions of CFCs that damage the ozone layer. This was amended in Kigali in 2016 and extended control to phaseout hydrofluorocarbons.

Since the institution of the Montreal Accord, the release of ozone-depleting chemicals has been reduced by 95%. Many nations still rely on CFCs, but work is being done to develop safe and effective substitutes.

 Ask Yourself...

Based on the health issues associated with ozone depletion, what can we do as a society to protect the ozone layer?

Global Warming ❗

Over the last several years, observations have shown that there has been a slow but steady rise in the Earth's average temperature. A 2013 report stated that most of the observed increase in the global average temperature since the mid-20th century is likely due to the observed increase in anthropogenic **greenhouse gas** concentrations. The three major gases are **carbon dioxide, methane,** and **nitrous oxide.** These gases absorb the infrared heat radiating from the Earth and thus heat the lower atmosphere.

Greenhouse Gas	Pre-Industrial Level	In 2016
Carbon dioxide, CO_2	280 ppm	400 ppm
Methane, CH_4	715 ppb	1,840 ppb
Nitrous oxide, N_2O	270 ppb	328 ppb

CO_2 and Temperature

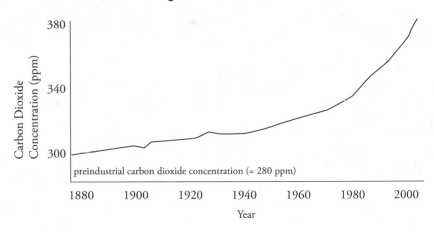

preindustrial carbon dioxide concentration (= 280 ppm)

Average Global Temperature, 1880–2000

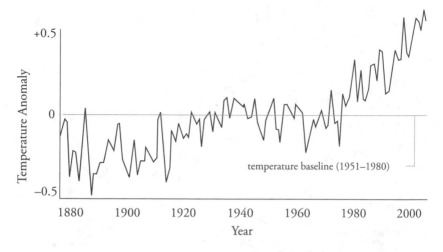

This warming is in addition to the normal warming of the atmosphere by the greenhouse effect, as shown in the following diagram.

The Greenhouse Effect

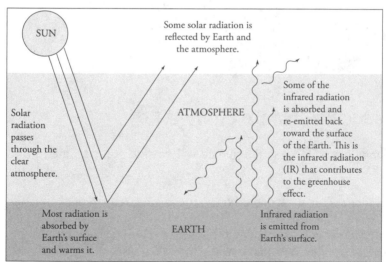

Effects of Climate Change: What's Coming? ❗

The increase in Earth's temperature will lead to a variety of changes:

Physical Changes
- Decrease of glaciers and ice sheets
- Continued rising of average ocean levels
- Changes in precipitation patterns
  - Wet areas will get more precipitation.
  - Dry areas will get less precipitation.
- Increase in frequency and duration of storms
- Increase in number of hot days
- Decrease in number of cold days

Changes to Biota
- Increased crop yields in cold environments
- Loss of croplands due to drought and higher temperatures
- Cold-tolerant species will need to migrate to cooler climates
- Heat-tolerant species (such as mosquitoes) may spread and invade new habitats
- Additional deaths from water and insect-borne diseases
- Commerce, transport, and coastal settlements may be disrupted by changes in ocean levels and storms
- Change in marine ecosystems and fishery productivity

Adaptations to the Warmer Climate
- Must occur at many levels of society
- Develop new technology
- Continue to reduce emissions from engines
- Legislative and behavioral changes needed
- Promote sustainable growth

Reducing Climate Change ❗

Adaptations to the warmer climate will need to occur at many levels of society:

- Technological—e.g., carbon sequestration, reduction of emissions from engines
- Behavioral—e.g., turning off lights to conserve electricity
- Policy—e.g., enacting new treaties and legislation

The promotion of sustainable growth will enhance the abilities of all societies to adapt to the new climate.

ASAP Environmental Science

Climate Change Laws and Treaties ❗

United Nations Framework Convention on Climate Change (UNFCCC)	Kyoto Protocol	Paris Agreement
International environmental treaty	International environmental treaty	International environmental treaty
Adopted in 1992	Adopted in 1997	Adopted in 2016
• Objective is to stabilize atmospheric greenhouse gas concentrations at a level that would prevent interference with climate	• Implemented to meet the objective of the UNFCCC • Commits state parties to reduce greenhouse gas emissions	• Required participating countries to reduce greenhouse gas emissions by a pledged percentage • The U.S. and China (who together account for over 40% of global greenhouse emissions) have both signed and ratified the Paris Agreement. • The U.S. had pledged to reduce its emissions by 26–28% below 2005 levels by 2025, but in 2017 President Trump began taking steps to withdraw from the agreement. • As of February 2018, the U.S. is the only country opposed to the Paris Agreement.

😎 When you meet someone new, start talking about global warming. It's a real ice-breaker.

Loss of Biodiversity ❗

Humans impact Earth through our interaction with animals. These interactions have led to loss of biodiversity—that is, a reduction in the variety of species across habitats and ecosystems.

Things we can do to reduce the rate of extinction include

- living sustainably and conserving resources
- making it illegal to trade in specific organisms
- implementing captive breeding programs
- conserving habitats by protecting large tracts of land from human activity

HIPPCO ❗

Use the acronym **HIPPCO** to memorize the causes of extinction.

- **Habitat destruction and/or fragmentation**
- **Invasives**
- **Population**
- **Pollution**
- **Climate change**
- **Overharvesting/overexploitation**

 Ask Yourself...

What are the potential problems that arise with an increased extinction rate among species? Consider the honey bee. How would the world in which we live change with the extinction of the honey bee?

Environmental Policy Acts ❗

Date	Agreement	What It Did
1970	National Environmental Policy Act (NEPA)	NEPA created the Council on Environmental Quality that resulted in the creation of the Environmental Protection Agency (EPA) from the consolidation of various environmental agencies. It also mandates that federal agencies prepare environmental impact statements.
1972	Marine Mammal Protection Act	This act protected marine mammals from falling below their optimum sustainable population levels.
1973	Endangered Species Act	The act prohibited the commerce of those species considered to be endangered or threatened.
1973	Convention on International Trade in Endangered Species of Wild Flora and Fauna (CITES)	This agreement bans the capture, exportation, and sale of endangered and threatened species.
1983	International Environmental Protection Act	This act authorized the president to assist countries in protecting and maintaining wildlife habitats, as well as in developing sound wildlife management and plant conservation programs. Special efforts should be made to establish and maintain wildlife sanctuaries, reserves, and parks; enact and enforce anti-poaching measures; and identify, study, and catalog animal and plant species, especially in tropical environments.
1990	Pollution Prevention Act	This act was designed to promote source reduction (stop pollution from being produced).
1990	Environmental Education Act	This established the Office of Environmental Education within the Environmental Protection Agency to develop and administer a federal environmental education program.

Sustainability 🔔

To environmentalists, **sustainability** usually means working in the biotic and abiotic environments in a way that ensures they are capable of surviving, thriving, and adapting into the future.

Public Policy 💬

The exploitation of public resources has been the motivation behind environmental policy. Policies made as a nation are usually fairly easy to enforce because they often have our collective best interests as a nation in mind. International policy, as is established through the United Nations (UN), is achievable and realistic only if the affected countries all cooperate.

National Policy	International Policy

• Easier to enforce than international policy
• State and local laws enforce environmental regulations, but federal law takes precedence if conflict arises
• Some states have stricter legislated controls than required by federal law
• Laws can be passed and enforced regionally due to geographic needs
	• Requires cooperation and collective decision-making
	• The UN can attempt to force countries to follow majority-rule mandates by
	  ○ restricting borrowing power through the World Bank
	  ○ withholding aid
	  ○ establishing trade rules
	• Effective policies have been ratified by treaties agreed to by participating governments
	• Impossible to punish countries that do not follow policies

---

Strictly speaking, *policy* is defined as a plan or course of action to influence and determine decisions, actions, and other matters.

Ask Yourself...

What would be the benefits of having a worldwide policy regarding the use of public resources? What would be the downfalls? Would such a policy ever be possible to achieve?

Political and Cultural Activism 💬

Mid-19th and Early 20th Centuries 💬

Shortly after the Civil War, as people continued to migrate to the West, it was realized that the United States did not have an endless supply of land or resources. During this period, there were a few early environmental activists.

Who?	What?
Henry David Thoreau (1817–1862)	• His book *Walden* describes his retreat from society and the quiet years that he spent living on Walden Pond studying nature. • Thoreau was a **transcendentalist**, or someone who believed that the most important reality is what is intuitive, rather than scientific. • Thoreau promoted the philosophy of living simply in harmony with nature.
George Perkins Marsh (1801–1882)	• His book *Man and Nature* helped the American public understand that there are limits to natural resources. • Marsh's plan for the conservation of resources is the basis for many of the resource conservation principles that we try to adhere to today. • Marsh also promoted the idea that the rise and fall of societies is directly linked to the sustainable use of resources.

Global Change

John Muir	• A nature preservationist who founded the **Sierra Club** in 1892
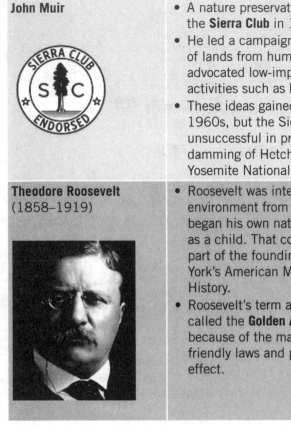	• He led a campaign for the protection of lands from human exploitation and advocated low-impact recreational activities such as hiking and camping. • These ideas gained traction in the 1960s, but the Sierra Club was unsuccessful in preventing the damming of Hotch Hetchy Valley in Yosemite National Park.
Theodore Roosevelt (1858–1919)	• Roosevelt was interested in the environment from an early age and even began his own natural history museum as a child. That collection became a part of the founding collection for New York's American Museum of Natural History. • Roosevelt's term as president has been called the **Golden Age of Conservation** because of the many environmentally friendly laws and policies he put into effect.

Theodore Roosevelt—The Conservation President

💬 Roosevelt increased area of national forest lands by 194,000,000 acres, established 150 new national forests, and inaugurated the first 51 bird reserves. Furthermore, he established 4 national games preserves, 5 national parks including the Grand Canyon, 7 conservation commissions, 18 national monuments (established under the 1906 Antiquities Act), and 24 reclamation projects.

1960s 🗨

Milestones affecting environmental policy in the 1960s include the following:

- **Apollo space missions**—Allowed Americans to see planet Earth from afar for the first time; popularized the term "spaceship Earth"
- ***The Population Bomb***—1968 book by Paul Ehrlich that warned of problems that would arise due to rapidly increasing human population
- **Nixon Presidency**—Initiated during this time were Earth Day (April 22, 1970); the National Environmental Policy Act (NEPA); and the creation of the Environmental Protection Agency (EPA)
- **Two key legislative actions**—Clean Air Act (1963), Clean Water Act (1972)

Green Taxes ❗

A worldwide discussion is taking place about how to move toward taxing environmentally detrimental activities like pollution. A new group of taxes called **green taxes** aim to

- generate revenue to correct past pollution damage and reduce future pollution
- use funds received from pollution taxes for restoration
- change behavior

💬 A **green tax shift** is a fiscal policy that lowers taxes on income, including wages and profit, and raises taxes on consumption, particularly the unsustainable consumption of nonrenewable resources. In this scheme, taxes serve as policy tools as well as a way to protect the environment.

 Ask Yourself...

How would the current world be different if the Clean Air and Clean Water Acts had never been passed? How do you think the population would be affected?

Market Permits !

Market permits are also being used somewhat successfully to encourage reduction in pollutants. Market permits are **cap and trade permits** and they work in this way: companies are allowed to buy permits that allow them to discharge a specific amount of substances into certain environmental outlets. If they can reduce their discharge, they are allowed to sell the remaining portion of their permit to another company. Economically speaking, it is to a company's advantage to reduce its discharge and sell the remainder of its allowable discharge to another company. The graph below shows the expected decrease in the number of permits available in the future and the expected increase in price.

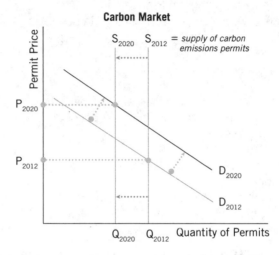

Carbon Market

Globalization ❗

Our world is becoming more and more interconnected. Aircraft can fly around the world in about 24 hours; we have instant communication worldwide via phones, television, and the Internet. This is called **globalization,** and it affects the economy, the environment, and society as a whole.

➕ Positive Effects of Globalization	➖ Negative Effects of Globalization
• New economic opportunities • Expanded access to information • Inter-society interaction • Increased access to resources	• Rising levels of air and water pollution • Rapid spread of emerging diseases • Increased levels of hazardous waste • Loss of biodiversity • Massive resource consumption

International Policy ❗

The World Bank uses loans to reduce poverty and to help foster improvements in biodiversity, environmental policies, and management of land, pollution, and water resources.

Through its environmental programs, the UN seeks to promote international cooperation, develop regional programs to promote sustainability, and to assess global, regional, and national environmental trends.

Ask Yourself...

Consider the following quote from former U.S. President Bill Clinton:

"No generation has had the opportunity, as we now have, to build a global economy that leaves no-one behind. It is a wonderful opportunity, but also a profound responsibility."

How will you make a difference?

Chapter 7 Summary

- How individuals manage their resource use and sustain the environmental quality of their biotic and abiotic environments dictates the future existence of the human species on Earth.
- Major domestic and international environmental policies have been created to influence and determine individual and corporate actions.
- Globalization has caused increased transportation pollution, unsustainable resource use, unequally distributed environmental issues, and poverty.

Are you an international student looking to study in the U.S.A.?

Visit us at
https://www.princetonreview.com/global